Harvey M. Bernstein & Andrew C. Lemer

SOLVING THE INNOVATION PUZZLE

Challenges Facing the U.S. Design & Construction Industry

Published by
ASCE Press
American Society of Civil Engineers
345 East 47th Street
New York, New York 10017-2398

ABSTRACT

This book provides a first-hand perspective of innovation in the design and construction industry. *Solving the Innovation Puzzle* outlines the obstacles to bringing innovative ideas into practice, including the influence of government policies and private-sector priorities. While many books only describe the problems at hand, this book provides possible solutions to help expedite the transfer of infrastructure-related innovation into practice, in order to assure an economically viable and competitive nation.

Library of Congress Cataloging-in-Publication Data

Bernstein, Harvey M.
    Solving the innovation puzzle : challenges facing the U.S. design and construction industry / by Harvey M. Bernstein and Andrew C. Lemer.
       p.   cm.
    ISBN 0-7844-0023-7
    1. Engineering design—United States. 2. Civil engineering—United States—Technological innovations. 3. Construction industry—United States—Technological innovations. I. Lemer, Andrew C. II. Title.
TA174.B447 1996   95-49030
620'.0042—dc20   CIP

CONTENTS

PREFACE

It is with some humility that we take on the design and construction industry. It is such a huge, sprawling, and intricately multi-faceted set of activities, institutions, and individuals that even to use a single word like "industry" seems inappropriate. We begin to form a vision of what it means when we are very young and excited by the prospects of massive equipment, deep holes, and the view from high places. Some of us never outgrow that excitement and enter into a partnership of builders, that extends unbroken from the dawn of human civilization. We have always had design and construction and they will undoubtedly continue for as long as our civilizations survive.

We admit to being zealots, but that is why we undertook to write this book. People outside the industry are often inclined to accept too easily the implication that thousands of firms and millions of workers can be comprehended as an "industry" and dismissed as hopelessly old fashioned. There is a tendency of those within the industry to focus on the details of their own situations and thereby miss the opportunities for concerted action for common benefits. We found in our work that what seemed obvious to us was not necessarily apparent to members of either group.

We hope to speak to both. We believe that technological innovation in design and construction can help us achieve the goal of bringing sustainable development and improved living standards to all the world's people. But this innovation will occur only if builders and bankers, owners and operators, users and neighbors, and all of the other stakeholders in our built environment agree that it should. If we can convince a few doubters and dawdlers, and motivate them to action, we will have achieved our aim.

We would like to thank our many colleagues who encouraged us and critiqued our thinking. We are especially grateful to those who reviewed our manuscript in its various stages and gave us suggestions for how to make improvements. We would also like to recognize the work of Sand Toler who created the cartoon sketches, on behalf of the Civil Engineering Research Foundation (CERF), which appear throughout the book. We thank our families too, for their tolerance of our time away and their willingness to hear the arguments several times over. While we certainly take full responsibility for our errors and failures, we cannot take all the credit if we succeed.

Finally, we thank the Board of Directors and staff of the Civil Engineering Research Foundation (CERF), who provided innumerable services and support. We could not have completed our work without the help of Liz Sigler of the CERF staff, whose abilities to assemble jumbled files and arcane notes into a credible manuscript—not once, but many times—were nothing short of miraculous.

—H.B. and A.L.

Chapter 1

THE DIFFICULT
WE DO IMMEDIATELY

On Monday, January 17, 1994, the Northridge earthquake shook the Los Angeles basin. Measuring 6.7 on the Richter scale, the quake toppled highways, ruptured gas lines and water mains, collapsed buildings, and disrupted the operations of the nation's second largest metropolitan area. Sixty-one people lost their lives beneath the rubble, 9,300 more were injured, and thousands suffered damage to their property. Hundreds of thousands had their lives disrupted for weeks. We watched in awe, along with much of the nation, as television brought into our homes the images of collapsed structures, flames, and billowing smoke.

Concrete parking garages display some of the worst damage from the Northridge quake. *Courtesy of S.K. Ghosh, Portland Cement Association.*

President Clinton offered immediate assistance of $379 million in disaster relief, but the immediate losses measurable in dollars were estimated by newspaper accounts to be as high as $30 billion, exceeding the more than $7 billion incurred less than five years earlier in the Loma Prieta earthquake that rocked San Francisco. The disruptions to business, major disruptions, real suffering, and psychological costs to people living in the area were much higher than even those staggering figures.

In the days that followed the earthquake, we read the news reports as armies of engineers and architects, construction workers, and others went to work to repair and rebuild. Thousands of employees, blocked from getting to their jobs by collapsed sections of the famous Los Angeles freeways, were forced to ride the region's new Metrolink rail transit system. Daily ridership increased from 950 to 22,000 passengers following the quake.[1] Within days, two new stations were built along existing tracks and service was extended 40 miles north to accommodate the thousands of citizens cut off from access to the city freeway system. Water was brought to areas isolated by broken mains. Slowly at first, and then with increasing speed, debris was cleared, overpasses were rebuilt, and pipes were repaired. People from private firms and public agencies worked together to get the job done and achieve a common goal. The news media reported in glowing terms on the rapidity with which the region's infrastructure was restored to working order.

The Kobe quake, also known as the Hyogo Prefecture quake, measuring 7.2 on the Richter scale, left Kobe in shambles.
© 1995, Michael Goodman.

As members of the professions that design and build the nation's constructed facilities —our buildings, bridges, water supplies, and waste treatment plants, the infrastructure on which we all depend—we observed the damage and subsequent recovery from these earthquakes with keen interest. Our particular concern for how our U.S. professions and industry are faring in a time of rapid global change made us examine the Northridge earthquake a little more analytically than we otherwise might have. Why did it seem so unusual that in a matter of a few months so much had been accomplished? What might have occurred if the new Metrolink transit system had not been there? Suppose the structures built in the last few years had been designed to meet older standards, set before we had learned more about the force of earthquakes.

As costly as the Northridge earthquake was, we were lucky when compared to Japan. Exactly one year after the Northridge quake, on January 17, 1995, the Kobe earthquake killed 5,250 people and devastated the densely settled industrial area. Experts estimated the damage at more than $150 billion. In a nation that history has given what is arguably the highest level of earthquake awareness in the world, what had gone wrong?

In the Northridge earthquake, many buildings and larger structures, built to meet the safety specifications of California's stringent design standards, weathered the shaking well. The standards themselves and many of the materials used in the construction represented innovations from older building practices.

There were still problems, nevertheless, as the damage figures attest. Engineers noted that rigid joints in steel-frame buildings cracked with alarming frequency. Some of these buildings were new. The brittle failures occurred in Kobe as well. We can congratulate ourselves for past innovations, but we cannot be complacent. There is more to learn.

The record-breaking speed of repair and recovery efforts, in both Kobe and southern California, also reflected innovation. Things got done quickly. There was a pressing need for action and people responded. Decisions that might normally have required weeks, or even months, of review by government agencies and others were made quickly and implemented. Resources were mobilized and work proceeded. Driven by need and freed from many of their usual restrictions the members of the design and construction industry were able to apply the ingenuity and innovation for which construction has been well known. The battalions of rebuilders might have adopted the motto of the Navy's Civil Engineering Corps (the proud SeaBees of wartime fame): "The difficult we do immediately; the impossible takes a little longer."

FEAR OF STICKING ONE'S NECK OUT

Unfortunately, the design and construction industry has come to be known more for its slowness to accept change and its apprehension toward implementing innovation. The procedures and safeguards we have created in the interest of public safety and competitive markets have deterred the U.S. design and construction industry from developing new products and services. Concerns about product liability, pressures to offer the lowest cost, lack of clearly defined quality standards, and a cyclical pattern of boom-or-bust in the marketplace are just a few of the reasons often given for why we have adopted low-risk-low-reward strategies in design and construction. Our system of building and managing the nation's constructed facilities seems to reward practitioners for conforming to past practices rather than innovation.

The Navajo Bridge, which now spans the Grand Canyon at Marble Canyon, AZ, is a unique steel-braced spandrel arch. *Photo by Richard Strange. Bridge design by Jerry Cannon for the Arizona Department of Transportation.*

The situation ought to attract more attention. The design and construction industry is at the heart of the U.S. economy. The various enterprises involved in design, new construction, renovation, and other construction-related activities, including equipment and materials manufacturing and supply, employ over 10 million people and account for roughly 13 percent of the nation's economic activity, as measured by our gross domestic product (GDP). Taken as a whole, design and construction comprises the nation's largest manufacturing activity! This industry is also a traditional source of entrepreneurial opportunity. While there are some very large and medium-sized firms, construction activity in the United States is dominated by over a million small businesses.

The design and construction industry has served our nation well. As the United States approaches the end of the 20th century, we can look back on a period of remarkable changes in our living patterns and landscape. For more than a century and a half, but especially in the decades since the end of the Second World War, rapid urban growth and suburban development have radically transformed the nation. This transformation has been accompanied by an extraordinary expansion of our investment in dams, buildings, highways, airports, and the myriad of other constructed facilities in what the National Science Foundation (NSF) termed as the civil infrastructure systems. Few people can fail to acknowledge the achievement this expansion represents. The United States today possesses a physical infrastructure of extraordinary scale and scope. This civil infrastructure supports virtually all elements of our society, and the people and businesses that have produced it comprise a major segment of our economy. History indicates that the growth, flourishing and decline of any civilization are closely mirrored by the life cycle and performance of its civil infrastructure.[2]

But even the most enthusiastic proponent of design and construction would acknowledge that the industry has problems. Because the industry and its products are so basic to our economy as a whole, these problems have very broad significance. In searching for solutions, we have discussed many of these problems in dozens of conferences, seminars, and meetings over the past several years.

The nation's built assets enable many of us to enjoy unprecedentedly high living standards, but we see all around us air pollution, traffic congestion, loss of open space, and other elements of environmental degradation. With the help of expanding transportation and water supply facilities, suburban "sprawl" has consumed land even faster than population has grown. Housed in often complex and inflexible factories and offices, businesses have evolved and then sometimes moved on, abandoning decaying inner city areas and producing sharp disparities in economic opportunity.

At the same time, the once-clear dominance of the U.S. construction industry and its technology in world markets has given way to increased international competition. These trends seem to threaten our livelihood and blur our views of what our national interests may be. They are not unique to design and construction. Rather, they exemplify changes going on in many segments of the nation's economy. Newspapers, television, internet and other media report daily on the continuing debate about national priorities, the roles of private enterprise and government, what rapidly emerging new technologies mean for people accustomed to using out-dated tools, and how we can best prepare our children for the new millennium.

Central to these trends and the debates on their causes is the idea of innovation. ***Innovation—new ideas, products, and ways to do things—expands our opportunities, increases our productivity, and improves our quality of life.*** Innovation has always been a mainstay of U.S. industrial leadership. Economists attribute a major share of the tremendous rise in the nation's standard of living in the half century since World War II to improvements in our productivity, driven by our technological innovation. But many people feel that over the past several years, something has happened in many segments of the U.S. economy to slow down the innovation and the productivity increase. Based on our experience in this country and abroad, we believe the

design and construction industry has been particularly hard hit. Because the industry and its products are so basic to our economy as a whole, the impact has very broad significance.

James Conant, a former president of Harvard University, once wrote "Behold the turtle: He only makes progress when he sticks his neck out." We have found that many members of the design and construction industry, as well as federal and local governments, have grown unwilling to stick their necks out to try new things. Perhaps even more importantly, we have found many of the industry's customers unwilling as well. The result has been certainly less dramatic than the Northridge earthquake, but as we will argue, potentially no less disastrous, both for the industry and the nation.

IMPROVING CONSTRUCTION: A NATIONAL CHALLENGE

This book is about increasing innovation in the U.S. design and construction industry, and why it is important to the nation. We are concerned not only with the processes of design and construction, the people, machines, and materials with which the nation is built, but with the products of construction as well. These products, the nation's buildings and physical infrastructure, are valuable assets, estimated at some $20 trillion, and are a legacy left to us by past generations.

These products include power plants, water works, roads, bridges, railroads, ports and airports, schools, hospitals, factories, communication systems, lifelines, and homes. This diverse collection of constructed facilities house, enable, and otherwise support virtually all other economic and social activity. When they fail to do their jobs well, everyone suffers. Getting the greatest possible return from our built assets is a challenge that should concern us all. Innovation in design and construction, encompassing operation, and use of the facilities as well, is then a challenge of national proportions. The industry itself and its customers must stick their necks out if we hope to get anywhere.

Design and construction has been an industry traditionally made up of many small businesses. The ease of entry has offered opportunity to many new entrepreneurs, making this a quintessential American enterprise. Today a relatively few large firms dominate the markets for major projects such as the building of new airport terminals, long pipelines, or skyscrapers. Yet, it is the thousands of small firms of architects, engineers, general and specialty construction contractors that are responsible for the design, construction and maintenance of a large portion of our built environment.

Enhancing innovation in the industry is then a challenge of consequence to many people. And because we all use and depend on the products of the industry, our premise here is that everyone has a role in meeting this challenge. We all can prosper from improving the processes and products of construction. Our goal in the following chapters is to show how we can meet the challenge.

In these chapters, we will argue in greater detail just how important our built assets are, not only to their designers and builders, but to everyone. We will show how innovation has improved both the processes and products of design and construction, but also argue that more is needed. The types of innovation we consider here are, of course, largely within the realm of the various professions that plan, design, and construct buildings, cities, and the networks of facilities that connect them. The individual entrepreneurs and large corporations that develop the materials and products these professionals use, however, are a much broader lot. In addition, government officials at municipal, state, and national levels play crucial roles in managing the public's capital, although taxpayers, ratepayers, and stockholders finance and own these assets. All of us have a stake in getting the most from our built assets.

We have observed that many U.S. design and construction firms are finding it difficult and unprofitable to be as innovative as they might like. New technologies developed by U.S. industry and academic institutions are being commercialized overseas. Our global competitors are becoming more successful, not because they are necessarily more inventive, but because they operate in a setting more conducive to spreading innovation in the marketplace.

Finally, we offer our prescription for change. In a regulatory climate that has grown increasingly restrictive over the past several decades, the nation's largest design and construction firms, acting with complete responsibility and social awareness, are often unable to face the financial consequences of potential litigation that may face the would-be innovator. Smaller firms cannot muster the resources to pursue the national or international markets that would make innovation profitable. Throughout the industry, innovation is stifled because the perceived risks overwhelm the potential benefits. We must change the situation, and that will take concerted national action, including federal and state policies promoting innovations and lifecycle performance of the civil infrastructures.

THE PUZZLE OF MAKING CHANGE

We liken the complexities of innovation in the design and construction industry to families and friends getting together to assemble a jigsaw puzzle. Everyone—designers, constructors, suppliers, owners, policy makers, and so on—must work together.

In our case, you start with a box that holds the pieces of several jigsaw puzzles that have gotten mixed together. We know generally what everyone is setting out to achieve, but we are not quite sure what the results will look like. Any of several pictures could emerge from the effort. We do not know for sure how many pieces are in the box.

Some people start looking for pieces with straight edges for the puzzle border, while others prefer to focus on the image, searching for areas of color or recognizable shapes. Some people will concentrate for long periods on the effort, while others come into the group to comment, and perhaps put in a few pieces, in between other tasks. Everyone tries to work quickly, both because the puzzle should be finished before the evening is over and because some competition often creeps into the group.

It helps if everyone can agree, at a relatively early stage, which of the several puzzles will be the goal for the evening. The activity of putting the puzzle together is enjoyable, but the real satisfaction comes in completing a picture. In fact, from time to time someone may slip a puzzle piece into his or her pocket, trying to be sure of the honor of inserting the last piece in the puzzle. Territorial disputes may even develop because someone wants to finish a part of the puzzle without any outside interference. Then squabbling could break out and the entire enterprise could fail. Once in a while, someone may have to be diverted to another portion of the puzzle, or the squabbling could escalate to frustration, and the pieces end up scattered around the room. If all goes well, by the time a puzzle is finished, everyone will have made some contribution, perhaps felt some frustration, and had some fun. Everyone feels a sense of accomplishment and looks forward to the next gathering.

As corporate executives, design professionals, project managers, craftsmen, legislators, insurance underwriters, government agency officials, researchers, lawyers, environmental activists, and users of facilities, we each play a role in putting together the innovation puzzle. Throughout this book we try to explain what these roles are and how we can do a better job working with each other in moving productive new ideas into practice.

Like the family gathering, the process of innovation takes many steps, which are not always clear or followed in a sequential manner. There may be lucky discoveries, but sometimes research is needed. Field demonstrations, pilot programs, government agency approvals, changes in standards and codes, test marketing, and gauging public reaction are a few of the steps in the innovation process that may have to be taken before the puzzle can be completed. We hope to shed some light on how to make the process more likely to yield satisfaction for everyone involved.

We believe that the design and construction industry, working in partnership with government, our academic institutions, and the public, can do a better job. Our ideas have largely grown out of our own experience as participants in what we hope will become a renaissance of design and construction innovation. We have had the opportunity over the past several years to observe design and construction activity in the United States and a number of other countries. We have been involved in critical studies of U.S. industry and government policy related to this industry. We see changes occurring in what is demanded of the industry, how the industry works and the products it delivers. We see forces at work that can direct this change in ways that will benefit us all, but these forces are still tentative, even fragile, and lacking in firm direction. We hope to influence the change.

We undertake in the following pages to put together our own picture of the puzzle. We will try to answer several key questions:

• Why does it matter how the U.S. design and construction industry fares in the global marketplace of the 21st century?

• What are the prospects for the technology of design and construction, that make us believe that a renaissance is possible?

• Does the U.S. have the incentives in place and the environment for innovation to benefit the design and construction industry?

• What can we do to ensure that the design and construction industry makes the greatest possible contribution to achieving the standards of living that we want for ourselves and for future generations?

We believe the last question calls for nothing less than a renaissance in design and construction. We use the term "renaissance," which means a return to youthful vigor, freshness, and productivity, because the design and construction industry was once the "high technology" leader in human enterprise. The great monuments of antiquity, the pyramids, the Great Wall, Europe's Gothic cathedrals, the factories, dams, and bridges of the Industrial Revolution, and our modern skyscrapers are highly visible examples of this leadership. Since the nation's birth, the design and construction industry professionals have been central to the United States' economic strength and high quality of life. As we face the problems of extending a high quality of life to all the world's people and sustaining that quality for future generations, the design and construction industry must renew its ability to make similarly bold contributions. We all must play a part in putting that picture together.

ENDNOTES

1. Don Phillips. "Quake Fails to Shake L.A. Commuters From the Driver's Seat." *Washington Post,* October 11, 1994, p. A3.

2. National Science Foundation (NSF). *Civil Infrastructure Systems Research: Strategic Issues.* K.P. Chong, Task Group Chairman. Washington, DC: NSF 93-5, 1993.

Chapter 2

DREAMING TODAY FOR TOMORROW

On a cold winter's day in 1852, John Roebling was stranded for a while in a ferry boat crossing New York's ice-clogged East River. The experience, we are told, may have been the inspiration for building the great bridge to Brooklyn that stands today as a landmark for millions of people. Roebling wrote of his plans in the preface of his 1867 Report to the New York Bridge Company: "The contemplated work, when constructed in accordance with my designs, will not

Brooklyn Bridge. *Courtesy of* ASCE.

only be the greatest Bridge in existence, but it will be the greatest engineering work of this continent, and of the age."[1] Roebling certainly had a dream!

At its opening in 1883, the Brooklyn Bridge seemed to many people symbolic of the great changes sweeping the country. The nation's urban population was burgeoning as new immigrants swept in from a troubled Europe. New technology was transforming the cities as great inventions were put into use—the railroads, gas and

then electric street lights, iron and steel skyscrapers, and the telephone, to name a few. The bridge itself embodied some of that change.
Steel cables from which the deck was suspended were woven by
innovative procedures of Roebling's design. By 1907, with two sub-
way lanes, two trolley tracks, and two lanes of horse-drawn carriages,
the bridge could carry some 426,000 people a day. Today, with six
lanes of autos and other vehicles, the bridge carries less than half that
number.[2]

The Brooklyn Bridge is part of what is arguably one of the
most well-developed infrastructures any nation has ever built. In
times past, the civilizations of Mesopotamia, Rome, the Incas, and
others have rested on extensive systems of roads and waterways, but
for sheer magnitude and diversity of technological components, the
United States of the late 20th century is unsurpassed. Building on the
work of our colonial forebears, we have spanned the continent,
bridged over and tunnelled beneath physical obstacles, and laid pipe,
cable, rail, concrete, and asphalt to connect ourselves through a vast
and complex multi-modal network. All this accomplishment is a demonstration of continuing innovation and a legacy that has served us
well.

Yet in 1899, only sixteen years after the Brooklyn Bridge
opened, the U.S. Patent Office's director, Charles H. Duell, urged
President McKinley to close the agency. Duell, with keen foresight,
judged that everything that could be invented had been invented. The
story is a favorite among students of technological progress.

SEEKING OUT THE INNOVATORS

The difference between Roebling and Duell was one of
imagination, often a key distinction between the entrepre-
neur—wherever he or she is found—and others. The difference is
crucial to innovation and to our puzzle. In the vast literature on
technological innovation and its economic consequences, creative
people willing to challenge conventional thinking and physical constraints often appear at the top of the list of factors needed for an

innovative society. We can find many more examples of such people, besides John Roebling, to support this observation.

The United States of the late 19th and early 20th centuries was truly an innovative society, an age and place of invention. Alexander Graham Bell, for example, transmitted a first few words over wires in his New Jersey laboratory in 1876. The first "long distance" lines were strung four years later, from Boston to Lowell, Massachusetts. By 1915, telephone wires stretched from New York to San Francisco.

The Empire State Building started a progression of innovations in the design and construction of taller buildings. *Courtesy of ASCE.*

Thomas Edison's experiments during the same period led to his incandescent bulb, electrical generator and distribution systems that came together when he closed the switches at his Pearl Street station in New York City. Commercial operations began in 1882, and by 1888, large electric light companies were found in many U.S. cities and in Europe. George Westinghouse's improvements on power generation and distribution had set the 60-cycle standard in this country by 1900.

As we have already remarked, the design and construction industry was then in the forefront of innovation. Our systems of sewers, water purification facilities and distribution lines, railroads, highways, and urban transit, along with telephone and electric power networks, expanded dramatically to serve a growing nation. A progression of taller and taller buildings began to rise with the Empire State Building in New York, the Sears Tower in Chicago, and then elsewhere. The high level of activity provided opportunities for thousands of innovators to make small improvements in design and construction technology that augmented the contributions of the major inventors we remember today.

But today the great discoveries are more likely to be made in other fields. Since the Wright brothers made their first powered flight in 1903, the aircraft industry has increased the speed and range of commercial aircraft manifold, while the price of air travel has plummeted. Doctors can use machines capable of substituting for malfunctioning kidneys and hearts to extend the lives of thousands of people. Entrepreneurs working in their garages found the way to put digital computers on desktops around the world.

Of course, innovation has by no means stopped in design and construction. For example, the steel and concrete produced now are stronger and more durable than materials used early in the century. Our efficiency in using energy has continued to increase as we learn more about technologies for generating and transmitting electricity, and do a better job of recovering heat that used to be wasted in the process. Over the years, however, a growing array of obstacles has been slowing the pace and scale of this innovation.

One of the more substantial of these obstacles, in our view, is a loss of excitement and imaginative thinking in design and construction. Within the industry, conservatism reigns as we seek to avoid risk by sticking with tried and true materials and methods. We are in an industry entrusted with the public's safety, so some degree of caution is certainly warranted in approaching innovation. But to outsiders who compare design and construction to other industries, we seem stodgy. Much of the basic technology on which we depend—e.g., steel and concrete structures, thermal and hydro-power electricity, petroleum-driven internal combustion engines—came into use in the 1800s or early years of this century. Research and development efforts, important sources of both inspiration and new technology, are modest by almost any measure, and even good results seldom attract the attention of the evening news.

What's worse, there is a snowball effect that works here: the lack of excitement and adequate pay makes it difficult to attract our brightest and most imaginative young people into design and construction. The cyclical—some would say sporadic—nature of the industry, with intense building booms in some places for several years, followed by similar periods of doldrums, offers further discouragement. Skilled labor and craftsmen as well as professionals cannot easily maintain stable lives in such a setting. As the most talented members of the design and construction industry age and retire, they are not replaced. Youths are attracted to other fields with more reliable prospects. Some of the resulting manpower deficit is filled by immigration of well educated students from economically less advanced countries. Some of them graduate from a U.S. university and then stay in this country, perpetuating a "brain-drain" in their native lands. Some of them, it is claimed, may replace workers who are less qualified or unwilling to work for the wages these immigrants accept. The social and political consequences of both cases are, at best, unattractive.

And so the design and construction industry has developed something of a reputation among its members, as well as with outsiders, for being slow to seek and accept innovation. While we will show that new ideas have in fact continued to come into practice, we do sometimes find it amazing that innovation is not altogether stifled by the Byzantine government regulations, labor practices that are outdated and sometimes bizarre to an outsider, and disjointed production relationships that the industry has evolved. At this early stage in our puzzle, let us say simply that the industry could use more imagination.

One of us had an opportunity some years ago to present the plans for a proposed new town and industrial development project to a large group of senior government officials in another country. The presentation was supported by beautiful watercolor paintings, several large models, plans and other drawings, and a cinematic presentation with eighteen slide projectors and soundtrack. All this and our short speeches were planned to convey our conception of how this new development might look, how it would work, and how it could be accomplished. One of the officials commented that we were helping them to "dream in the daytime," and then proceeded in the following months to put the plans into action. Perhaps all of us in the design and construction industry could do a bit more dreaming in the daytime.

But we need something else as well. While it may seem as though we are splitting hairs, we think it is important to make a distinction between dreaming up a new product or process—the invention—and subsequent innovation. It is putting new ideas (i.e., new products or procedures) into practice that determines innovation. People must dream, but they also must act. Often, two

Images like this artist's rendering of a new town help people imagine what is possible and thereby help assure that plans will be realized.

different types of people are needed to accomplish innovation. Futurist Joel Barker terms these two types of people as "paradigm shifters" versus "paradigm pioneers." While the "paradigm shifter" formulates a new concept or process as a necessary precondition for innovation, the "paradigm pioneer" must make the innovation effective and often reaps the rewards. John Roebling is one of those unique individuals who we can classify as both a "paradigm shifter" and a "paradigm pioneer"; he not only had the vision to create the concept of the Brooklyn Bridge, but he also had the drive to see it built.

The example of video cassette recorders (VCR) is often used to illustrate how U.S. industry as a whole seems frequently to fail in the follow-through required to become "paradigm pioneers." The basic technology originated in U.S. industry laboratories, but Japanese manufacturers brought the VCR into living rooms across the nation. Some industries seem to be learning the lesson better than others, and we think design and construction could do better.

THE PROMISE OF NEW TECHNOLOGY

Now is a good time to dream. We face a number of vexing problems that the design and construction industry could help solve. Environmental degradation has accompanied economic development. Human hunger and disease are concentrated where development has been slow. At the same time, the overall rate of technological progress remains breathtakingly high, with new developments in materials, electronics, biological and medical sciences, mathematics, and other fields emerging almost daily. We do not intend to propose our vision of a new world, but we cannot help asking ourselves the **"What if...?"** kinds of questions about what we find now in the world around us.

When Pat Choate and Susan Walter published *America in Ruins* in 1981, for example, they warned that our public facilities were wearing out faster than they were being replaced. Subsequent studies made dire assessments as well, but the public's reaction might be best characterized for the most part as "out of sight, out of mind." The subsequent decade of study and national debate failed to yield much in the way of a public policy response to a problem that most people in the design and construction industry know all too well. The problem is not too little replacement, but rather the failure to maintain what we have.

This is a problem that most of us face daily with our automobiles, our homes, and even our personal health; we tend to ignore the possibility that something might go wrong until something actually does. Check-ups and preventive maintenance seem like a waste of time if nothing is found to be wrong. Sometimes we simply would rather not know that something might not be right. For buildings, highways, and other large facilities, we enjoy cutting the ribbon when new construction is completed, but seldom celebrate maintenance well done. We then fail to provide the funds or take the time for maintenance.

Whatever the reason, we lose millions of dollars annually, not to mention countless hours and unmeasurable aggravation, from having to make major repairs where problems might have been kept manageable or prevented altogether. We respond to large potholes in the highway, leaking roofs, or burst water mains, but the damage has been done.

What if we had reliable tests or indicators of the current condition of all kinds of constructed facilities that made it very difficult to ignore the need for maintenance? Forty years of research and development have brought us close to having such indicators for highway pavements. Many state departments of transportation have begun to use "pavement management systems" that keep track of key data and help predict where and how much maintenance is likely to be needed. Regular inspections collect much of these data, using equipment developed for the task. Crews fill cracks or overlay worn pavements with a new riding surface, according to schedules these management systems produce. The whole procedure helps the agency treat its highway network the way an airline treats its airplanes, fixing things before they fail. These agencies are finding, for the most part, that this approach can save money as well as provide better service to the public.

What if we managed all our facilities that way? Take for instance the Kingston Bridge in Glasgow, Scotland. Built in the 1970s, the Kingston Bridge is one of the most heavily traveled bridges in Europe, carrying over 154,000 vehicles a day.[3] Underestimation of travel flows and flaws in the design, materials, and construction methods resulted in deficient bridge performance. In the late 1980s structural defects were detected that prompted the implementation of one of the most extensive structural monitoring systems in the world.

The Kingston Bridge monitoring system is complicated by the fact that the bridge is a dynamic structure with significant changes in its dimensions both daily and seasonally. Changes in weather conditions dramatically affect bridge performance. In order to monitor the bridge performance effectively, a variety of instrumentation devices have been installed. Bridge deck temperatures are continuously monitored by thermometers installed at different depths in the concrete, while meteorological stations monitor air temperature, relative humidity, wind speed, wind direction, and solar radiation. Global movements are surveyed once per month, while horizontal and vertical move-

The ten-lane Kingston Bridge, an integral part of Glasgow's highway system, is instrumented to give maintenance workers data that helps them plan what needs to be done to keep the roads open and safe. *Courtesy of Donald Carruthers, Strathclyde Regional Council.*

ments within the bridge are continuously monitored. All performance data are transmitted via a dedicated telephone line to a data processing unit off-site. The engineer thus has constant access to real-time data without performing on-site inspections.

Such monitoring systems can extend the life of a structure by providing crucial performance data for making maintenance decisions. It is too soon to know what the total savings will be in the bridge's maintenance and repair costs, but there is little question that the bridge's users are safer because of the monitoring system. As an extra dividend, the data help make better decisions about design, construction and maintenance of other facilities as well.

What if the scientists and engineers, who tailor materials from the atomic scale upwards, turned their attention to such seemingly mundane products as asphalt and Portland cement concrete? There already is a good bit of research going on to enhance the performance of these materials, but most of it is focused well above atomic or molecular scales.

The payoffs of improvements in these materials could be enormous. More than 500 million tons of concrete, for example, are produced annually in the U.S. for construction activities. That represents about two tons per person. Today's typical concrete compressive strength, i.e., the stress at which failure is predicted to occur, is 4,000 to 5,000 pounds per square inch (psi). That is more than twice as strong as the typical concrete produced early in the century. But much greater strengths are possible.

Taisei Corporation uses its automated construction system (T-UP) to build the 34-story Mitsubishi headquarters in Yokohama. With this system the sheltered factory floor allowed robots to construct the structure and skin of the building right on-site. As each floor was finished the factory was then jacked upward to begin work on the next floor while work continued on the interior of the completed floors. Using this technology, the steel erection averaged three days per floor. Overall construction of the Mitsubishi building took 24 months.
Courtesy of the Taisei Corporation.

Strengths in excess of 10,000 psi are routinely being produced in tests and demonstrations, but practical applications of this higher strength material remain rare because standard design manuals and building code regulations have not yet been developed. Laboratory experiments in France are producing strengths of 100,000 psi and higher. The value of such strength is multiplied because, for most concrete structures, the weight of the structure itself often represents the heaviest load to be supported. An experimental bridge actually constructed with 8,700 psi concrete in France, for example, used 30 percent less material than would have normally been needed. Suppose we could cut the materials use even further. Construction time and energy requirements would be reduced as well.

What if the cost of aluminum could be substantially reduced? While aluminum alloys offer light weight and excellent corrosion resistance, their initial cost compared to conventional steel and concrete structures keeps them from being used in many construction applications. The light weight and low maintenance of aluminum structural members could cut construction time for large facilities by as much as one-third, and permit pre-assembly of large structural sections, giving tighter quality control.

What if robots or remote-controlled construction equipment could take over the more dangerous or tedious jobs of construction, such as placing steel in tall buildings, excavating tunnels, or cleaning storage tanks? Workers' safety would be improved, which would also translate into savings in construction time and costs. Workers' productivity would rise as well. Japan, a country very conscious of productivity due to severe skilled labor shortages, is taking the lead in using automation and robotics to transform the building site into a construction factory.

What if bio-engineering could produce new varieties of wood suitable for construction? Many people feel wood is the closest to being a "sustainable" building material, but over-harvesting, environmental regulations, and increasing world-wide demand have caused wood prices to rise steadily as the quality and quantity of supplies decline. U.S. homebuilders are finding they must search for alternative mate-

Corporations in Japan are currently considering a future where vast domes will be erected over dense urban centers, providing climate and energy control and reducing the cost of buildings. *Courtesy of the Taisei Corporation.*

rials if they are to keep the costs of new and rehabilitated houses at levels people can afford.

What if we had non-polluting motor vehicles? **What if** we had electronic controls that would maintain steady speeds and safe distances between vehicles on the highways, and thereby improve both safety and traffic conditions during the periods of peak travel demand? **What if** we had an effective way to meet our electricity demands with solar energy. **What if...?**

What if...? We could go on and on. Our point is that there are many improvements to be made in design and construction, and we all can imagine what some of them might be. Most of the improvements we have suggested do not require that great of an advance in technology, although one could easily imagine much more dramatic changes. Visionary designer R. Buckminster Fuller, for example, dreamed more than thirty years ago of domed cities, able to control their pollution and reduce their aggregate consumption of energy. While these ideas seemed 'far-fetched' at the time of their conception, the con-

Manhattan under a dome as envisioned by R. Buckminster Fuller 35 years ago. ©1960 Allegra Fuller Snyder. *Courtesy of the Buckminster Fuller Institute, Santa Barbara.*

cept of domed urban centers is currently being explored in Japan as possible solutions to contemporary problems. Very large "mega-mall" shopping centers are not so different in concept, although their scale is much smaller. Proposals have been made for high-speed, trans-continental trains, operating in tunnels; mile-high buildings that would shelter many people and businesses without using so much land, while totally recycling all wastes; and floating cities devoted to environmentally benign farming of marine life, relying on power from the sun and the oceans. Such visions could become possible as advances are made in technology and applied in an environment conducive to innovation.

Experts often classify sources of such innovations into two categories. "Technology push" occurs when a new technology is invented that people then use in unanticipated ways to improve their lives. Nobody could have imagined, for example, the impact the telephone would have. Many people at first could not understand why anyone would want to use the telephone at all. Historians tell the story of bureaucrats in the 19th-century German post office who initially would not touch the instrument, viewing it a menial tool, beneath a gentleman's dignity. As late as 1932, there was a wooden telephone booth outside the President's Oval Office in the White House, and not until Franklin D. Roosevelt did a U.S. President actually allow the telephone into the office.[4]

The other primary source of innovation is "demand pull." A strong need or demand for a particular service or product draws people to search for a way to provide it. Needs and desires may stem from a problem to be solved, a chance for profit, concern for people's suffering, the wish to protect a place, or other motivations. The old saying that "necessity is the mother of invention" reflects an understanding of the importance of this "demand-pull."

Typhoid, cholera, and other illnesses caused by contaminated water were widespread in the 19th century, for example. Sand filtration of the urban water supplies spread rapidly in the latter part of the century, after systems in Lawrence, Massachusetts, and Poughkeepsie, New York, dramatically reduced the typhoid death rates.[5] Chlorination was a later innovation that was similarly "pulled" by demand.

We see both forces at work in the design and construction industry, but lately the pull of demand is especially strong. Increased understanding of the environmental consequences of our past construction practices and a burgeoning number of specific environmental regulations, for example, have encouraged businesses to find new ways to do things. Whole new lines of business have arisen, dealing with storage of toxic materials, enhancing the stability and strength of soils and rock, and measuring minute quantities of contaminants in air and water. Recurring disasters from earthquakes, major storms, and flooding, increasingly damaging as our population grows, have encouraged development of new structural designs and construction materials able to withstand these natural forces. New markets are emerging for glass- and carbon-fiber reinforcing for damaged structures, devices that isolate buildings from the shaking of their foundations, and other kinds of devices that allow structures to keep internal stresses within allowable limits by compensating for the earthquake's shaking.

We will examine in our next chapter some of the innovations that have occurred in recent decades and the reasons they have succeeded in meeting the needs of the industry and society. We are especially excited, however, by the subject that follows afterward, the prospects we see for the design and construction industry to be pushed to innovation by advances in other fields and more basic science and technology.

THINKING GLOBALLY BUT ACTING LOCALLY

Whether the design and construction industry is pulled or pushed to innovate, we have noticed in our travels that innovations seem to be emerging more quickly and strongly in other countries. In our visits to construction companies, design firms, and product manufacturers in the industrialized countries of Europe and Asia, we saw first hand that new technologies are being developed for the built environment. We have seen similar work going on in this country, but generally on a more modest scale and entering practice at a slower pace. If, as we believe, design and construction and the industry's products underlie economic productivity, then we as a nation may find ourselves either increasingly dependent on imported technology to maintain high returns on our built assets, or simply falling behind in our abilities to attract capable people and investment and to deliver to our population the increasingly higher standards of living we have come to expect.

As our examples here and in the next chapter illustrate, the United States has historically been a very successful producer of new design and construction technology. We have found in our travels to Japan and Western Europe that the U.S. design and construction industry continues to be a good source of new ideas, although we no longer have such a lead as we once did. Others have found the United States to be a leader in a few areas, but that our competition is rapidly closing the gap.[6] Almost everywhere, funding of new technology has gotten more difficult for many companies to justify because the immediate payoffs are sometimes difficult to demonstrate. Almost anywhere, emerging new technology may be so advanced that many people cannot envision what its value might be. And there are other problems of the innovation puzzle that we will discuss as well.

While we and others are concerned about the position of the U. S. design and construction industry in the global marketplace, that is not a primary issue here. We believe everyone in the United States and elsewhere, should be able to gain from solving the innovation puzzle. New technologies do not become innovations unless they are put into practice. Discovering the technology and perfecting it are only the first steps in an often arduous innovation process. And only some of the benefits of innovation are felt by the industry itself. Users, owners, and neighbors of facilities—the design and construction industry's customers—enjoy the improved services, greater reliability, and lower costs that innovations bring. These facilities—the industry's products—are valuable assets that we inherit from previous generations and pass on to our children. Everyone has a stake in getting the greatest possible return on these assets.

The benefits of innovation are realized through actions that are taken at local levels, by owners, users, and neighbors, each time a facility is planned and designed, constructed, used and maintained, renovated, reused, demolished, or abandoned. These stages of a facility's development, from initial planning to reuse or disposal, comprise what is generally termed the facility's life cycle. Innovation can be introduced at every step along the way. Our purpose here is not to predict technology or create a specific picture of the future, but we do believe the design and construction industry worldwide has a number of opportunities that will have far-reaching effects in the marketplace:

● *New information and analysis tools* that will improve the quality of our decisions about siting, scale, and functional character of new and renovated facilities

● *New materials* used in design and construction, with higher strength, wear- and corrosion-resistance, requiring less energy in their production and fabrication, and derived from renewable sources

● *New design configurations* that will provide healthier, safer, and more productive service and shelter

● *New designs and methods for production* that will shorten the times from conception to commissioning and consequently lower costs

● *New equipment and procedures* that will enhance worker safety, save energy in operations, and reduce pollution and waste

● *New management tools* to forecast the behavior of facilities over the full course of their service lives, leading to better maintenance and reduced costs overall

● *New methods for protecting our environment* leading to environmentally friendly facilities

● *New systems for materials handling and processing* to facilitate construction and recycling of all elements of old structures that are no longer needed

These are some of the results we can imagine the design and construction industry achieving, if we can solve the innovation puzzle. If we are to succeed, we in the industry need to do better at "dreaming in the daytime" and then making our dreams realities. We believe our past experience offers lessons about how we can do better. We believe that what is happening in other countries offers lessons as well. Let us now explain what we mean.

ENDNOTES

1. Alan Tractenberg. <u>Brooklyn Bridge: Fact and Symbol</u>. Chicago: University of Chicago Press, 1965.

2. Clark Wieman. "Road Work Ahead: How to Solve the Infrastructure Crisis." *Technology Review*, January 1993, pp. 42-48.

3. P.H. Milne, D. Carruthers, and A. McGown. "Monitoring Movements of the Kingston Bridge, Glasgow, by Land Survey Techniques." *Civil Engineering Surveyor*, November 1992, pp. 22-24.

4. "When Communication System Fails, Civilization's Lifeline Seems to Snap Service Economy and a Sense of 'Connectedness' Depend on Phones." *The Washington Post*, June 27, 1991, FINAL Edition. By: Curt Suplee, *Washington Post* A SECTION, p. a25. Pool, Ithiel de Sola, editor. 1977. "The Social Impact of the Telephone." MIT Bicentennial Studies 1. Cambridge, MA: MIT Press.

5. Eugene P. Moehring. "Public Works and Urban History: Recent Trends and New Directions." *Essays in Public Works History, No. 13*. Chicago: Public Works Historical Society, 1982.

6. Japanese Technology Evaluation Center (JTEC). *JTEC Panel Report on Construction Technologies in Japan.* Baltimore: Loyola College, 1991; and Civil Engineering Research Foundation European Research Task Force. *Constructed and Civil Infrastructure Systems R&D: A European Perspective.* Washington, DC: Civil Engineering Research Foundation, 1994.

Chapter 3

SPEAKING UP FOR INNOVATION

What do we really mean by the "design and construction" industry? So far, we have been writing rather generally about the activities of design and construction, and about infrastructure, buildings, and related products of these activities. We have been writing from the perspective of two people who have chosen to spend much of their working lives dealing with such matters, so we at least think we know what we are dealing with.

When it comes to putting down facts, we have found in our a surprising lack of detailed statistics or definitive analysis of design and construction as an industry. There have been a few widely spaced studies that sought to explain "problems," for example why construction productivity is low, what role construction plays in economic development, or why U.S. construction no longer leads in international markets.[1] Despite the years separating

these studies, the reports often sound remarkably similar. Perhaps the problem has been one of perceptions, those of outsiders attempting to study the unique practices and relationships found in the design and construction industry. Economist P. J. Cassimatis, who authored a study published in 1969 by the National Industrial Conference Board, a non-profit, industry-sponsored research organization, found that the construction engineering literature made it readily apparent that substantial gains in construction productivity had occurred over the previous decades, but "this apparent progress has yet to be revealed by economists."[2]

Perhaps the problem is that the design and construction industry has kept too much to itself in revealing its achievements. Henry Kaiser, founder of the construction company and a conglomeration of industries, many still bearing his name, was reportedly fond of saying "If your work speaks for itself, keep your mouth shut!" While we do believe that actions speak louder than words, not everyone has the opportunity to observe first-hand the results of the design and construction industry's work. Maybe it is time for this industry to speak out more aggressively for itself and its achievements.

DEFINING DESIGN AND CONSTRUCTION

Viewed in terms of the numbers, the industry is impressive. New construction in the United States typically employs six million people and produces an annual output estimated by the U.S. Department of Commerce to have been approximately $470 billion in 1993. That figure represents eight percent of the U.S. gross domestic product (GDP). Even though the dollar amount of annual output has increased over the years, the percentage of GDP has declined since 1966 when new construction accounted for 12 percent of the GDP.[3] If we consider the large repair and retrofit segments of construction activity, with the manufacturers and suppliers who provide materials and equipment, we estimate the industry is responsible for approximately 13 percent of all U.S. economic activity.

Moreover, construction influences the productivity of other sectors of the economy because of its core role in the nation's infrastructure. For example, the Office of Technology Assessment (OTA) has estimated that the multiplier effect from construction spending, the total dollars eventually spent throughout the economy for every dollar spent on construction, is nearly three![4] And as demonstrated in the work of a number of noted economists, such as Alicia Munnell, David Aschauer, and Roy Jorgenson, productivity growth closely follows the rate of investment in public infrastructure, one of the industry's principal segments.[5]

If we add up the work of planning and design; manufacturing of construction equipment, construction materials, heating, lighting, and other equipment and products used in buildings; facilities operations and maintenance activities; financial services; and the myriad of other busi-nesses that depend on the produc-tion and use of highways, buildings, power plants, and other fa-cilities, we feel very safe in asserting that innova-tion in design and construc-tion can influ-ence rather directly no

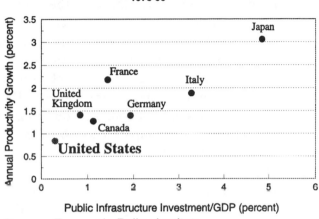

**Public Investment and Productivity in the G-7 Countries
1978-90**

Source: Economic Policy Institute.

less than one-sixth of the nation's economy. If we then consider how much we all rely on our homes, offices, rails, roads, powerlines, pipe-lines, and more—the ultimate products of the design and construction industry and its workers—it is hard to imagine any activity short of growing food or protecting water resources that is more crucially important to our lives and well-being.

One of the key characteristics of design and construction distinguishes the industry from most other segments of the economy and may account for its apparent neglect in the scholarly literature. Most of the industry's production occurs as a series of individual, unique projects. With the exception of "manufactured" housing and a few other applications of pre-fabrication, there is little opportunity for using mass-production methods. While such methods are em-ployed in production of raw materials and building components (e.g., plumbing and electrical fixtures, tiles, cement, and lumber), there is an extraordinarily high level of custom design and manufacturing even in the inputs to construction. While we think it is an exaggeration, some observers claim the Industrial Revolution bypassed design and construction.

The industry's modest efforts to innovate are a sharp contrast to its huge role in the nation's economy and welfare. In comparison to other industries, design and construction are lagging. While research and development (R&D) is not the only source of innovation, evidence from other industries suggests it is at least an important indicator of the likelihood that new ideas are being sought. For the design and construction industry, the effort is small indeed. A recent survey by the Civil Engineering Research Foundation confirms earlier studies that have found average annual R&D spending by all segments of the design and construction industry averages 0.5 percent of total revenues, compared to 3.4 percent for a composite of major U.S. industries.[6] In a more limited study of U.S. parent companies with foreign affiliates, the U.S. Department of Commerce found that construction companies in 1989 devoted an amount to R&D equivalent to $270 per employee, as compared to more than $3,190 per employee spent by all U.S. non-bank parent companies. Employees engaged in R&D work represented only 0.8 percent of total construction company employment, compared with 3.4 percent of U.S. parent companies in all industries.[7]

Such statistics are strong circumstantial evidence supporting the view that the design and construction industry is not particularly inclined to seek innovation. Various studies have explored this view and found a number of plausible explanations for why this might be the case.[8]

Most frequently cited is the highly fragmented character of the U.S. design and construction industry. While there are a few large firms, the industry is characterized by hundreds of thousands of small "mom & pop" operations. Approximately 85 percent of all design and construction firms employ fewer than ten persons, and the majority of them have no more than four employees. Less than one-tenth of one percent of all design and construction firms employ 500 or more persons.[9] Small firms have neither the resources to undertake substantial R&D programs nor the market potential to derive sufficient benefits from developing proprietary new technology.

To make matters worse, companies involved in construction, and to some extent in design as well, tend to operate on extremely thin profit margins. The resources for consciously pursuing innovation—whether through R&D, trial applications of new materials or processing techniques, or other means—are often relatively limited, even in larger companies. As we will describe later, there are also many other factors unique to design and construction—e.g., fear of litigation and facility-owner emphasis on low construction cost regardless of longer-term consequences—that tend to exaggerate the impact of these structural characteristics that weaken the industry's commitment to innovation.

Consider also the people involved in design and construction. Many of them are craftsmen and professionals in the traditional and truest sense of those terms. They often undergo years of education, training, and apprenticeship to enter their fields. They learn through experience, continually enhancing their skills and judgement. Their education, both formal and on-the-job, is recognized as a primary factor contributing to higher productiv-

Courtesy of ASCE.

ity. It is this more productive workforce that routinely assumes responsibility for millions of dollars of assets and the physical comfort and safety of thousands of people. Many of them routinely expose themselves to dangerous conditions in the course of their work.

For these people, new technology can be more a threat than an attraction: they often perceive that the risks of departing from tried-and-true ways of doing things far outweigh the potential benefits achievable if the new idea works. Addressing their concerns and changing that perception are two more pieces of the innovation puzzle.

DEFINING INNOVATION'S BENEFITS

Let us try to be more explicit about the benefits of innovation in the design and construction industry. As we already noted, innovation entails both invention and application. Experts often identify three primary stages in the process: (a) production of new technology or information, (b) its transfer to people who can use it, and (c) its subsequent application to solve problems, to see the world in a new way, or to enhance our efficiency, effectiveness, or quality of life. Innovation is important not for its own sake, but rather for the benefits it can bring to the individuals, organizations, and societies that use it, and it is only after reaching the third stage that benefits are realized.

In the design and construction industry, innovation can mean a new product or production process, the substitution of a less expensive, more durable material in an otherwise unaltered product, or the reorganization of production, internal functions, or distribution arrangements, leading to increased efficiency, better service, environmentally friendly results, or lower costs. In this context, "product" and "process" cover a

Sunshine Skyway Bridge in Tampa Bay, FL. *Courtesy of ASCE.*

broad range of possibilities, from a bag of cement to an entire bridge, from the quarrying of crushed stone for concrete to the design of sophisticated electronic controls for an electric power generating plant.

Despite our concern about the current rate of innovation, the design and construction industry has introduced quite a few new technologies and ideas over the past several decades. Some of the changes are obvious. People who drive across new cable-stayed bridges like the Sunshine Skyway between St. Petersburg and Tampa, Florida, cannot help but be impressed by the "lightness" of the structure in comparison to the steel frames and massive concrete of the past. The cable-stayed suspension system allows concrete bridges to span much greater distances than more traditional concrete designs. The innovation yields an aesthetically pleasing work of art that is also more cost-effective to construct.

Similarly obvious, but more likely now to be taken for granted, are the huge cranes and reusable steel formwork for casting concrete that assist the constructors of tall buildings to do their jobs more quickly and with less disturbance to the neighbors. The cranes tower above the skyline as a symbol of growth and prosperity in an urban area.

Less obvious—because they are typically not visible to the naked eye—are the growing number of "smart" devices and

Cranes and other heavy machinery are rarely noticed in our construction-oriented society. *Courtesy of ENR, Dec. 19, 1994.*

monitoring systems that diagnose and keep track of the condition of a constructed facility. We already mentioned the case of Glasgow's busy Kingston Bridge. When the structural engineer for New York's

Citicorp building realized that certain joint design details of the building put the structure at risk in high winds, he and his colleagues had sensors and wiring installed that monitored the building's motion in much the same way as a hospital patient is monitored in intensive care. The continuous readout of data could be watched at one of the engineers' offices, several blocks away, until repairs had been completed.[10]

Other design and construction innovations supply us with cleaner and more comfortable indoor air, energy-efficient lighting, and smoother-riding road surfaces. Machines now dig tunnels that formerly required dangerous explosives and hand-labor. We could go on and on, but have chosen to focus on the few examples that follow. We feel these innovations have brought substantial benefit—in terms of durability, safety, and a range of other factors, both qualitative and quantitative—and illustrate useful lessons in what it takes to solve the innovation puzzle.

COMPUTER-AIDED DESIGN AND CONSTRUCTION

As is the case in many industries, computers today are being used in virtually every aspect of design and construction. In the 1960s, early progress in computer-aided design was spurred by the development of the COGO system, named for its ability to *co*ordinate *ge*ometry, the basis for surveying land, siting structures, and locating where the various parts of a constructed facility should be placed. Developed initially at the Massachusetts Institute of Technology, the system in practice helped highway engineers locate new routes and structural engineers analyze building frames. This effective and—for its time—user-friendly computer program package was a basis for the development of other computer-based analysis tools for soil and rock foundations and embankments, steel and concrete structures, highway construction planning, and other applications. These tools required large computers and programming sophistication that limited their use to those professionals who specialized and firms that were large enough to justify the investment of time and money needed to gain access to a system. They were nevertheless a substantial innovation that advanced the industry.

General-purpose computer-aided design (CAD) systems began to appear in the 1970s, offering more effective data management and faster updating of the huge number of drawings required for the construction of a large facility, but they too required expensive hardware and specialist users. It was only when powerful and relatively inexpensive desktop computers were developed—the PC (personal computer) and distributed workstations—that CAD usage became the norm. Lower-cost hardware was followed by easier-to-use software, and the total investment required to begin using CAD came down to levels that virtually any engineer, architect, builder, or supplier could afford. Now, in the 1990s, powerful portable computers—the "laptops"—are light enough to travel around the world with a full set of working drawings in storage, and durable enough to go on the construction site. And just as lower-cost and more powerful computers have enabled the innovation in data availability, so has the development of "object-oriented" programming given rise to software that more closely matches the ways designers work.

CAD (and its computer-based cousin, the geographic information system, or GIS) is doing more than simply saving on the labor required to make design drawings.[11] It is changing in very basic ways how business is done. Perhaps the most widely reported example of what is happening is the production of the new Boeing 777 aircraft. The entire design was created on a CAD system, without the traditional use of costly full-scale mockups. The time required for Boeing and its industrial associates to go from initial conception to final design was cut dramatically. Automobile manufacturers have developed the same capability, linking their designers, allowing them to work collaboratively even when they are scattered in several cities; and similar methods are being applied in facility design and construction. Gensler and Associates, the San Franscisco-based architectural firm responsible for the implementation of the design of stores for such retail giants as the Gap and Banana Republic, is able to customize standard designs to fit a particular setting, bring the construction team together, and order all fixtures and equipment using its CAD system. Firms like Gensler are able to oversee corporate "rollouts" of hundreds of stores being developed simultaneously in cities around the world.

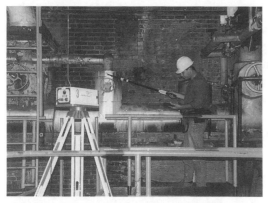

The Odyssey™ System allows for real-time position measurement. The system uses portable receivers that calculate position based on signals from laser transmitters (above). Odyssey can be linked to computer-aided design (CAD) in real time, thus allowing direct capture of positions into "as-built" CAD models, or site layout directly from designs created in CAD (below). *Courtesy of Spatial Positioning Systems, Inc., Reston, VA.*

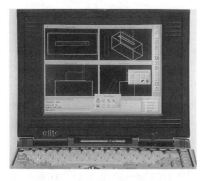

Stone & Webster and other firms that build large and complex electricity generation and chemical processing facilities, for example, now use 3-D CAD systems to check for interference among structural, electrical, and mechanical system components, that could prevent construction of a design. They also use the system to simulate construction operations, assuring that heavy equipment can get into the spaces where it must operate and out again when the job is finished. In the past, such applications were limited to the most costly and complex of facilities, but they are increasingly common as the cost of computer power and software continue to decline.

The applications are extending into the construction operation as well, adding CAD capability for on-site construction operations to the scheduling and labor- and materials-management tools routinely used by leading constructors. The Odyssey™ System, created by Spatial Positioning

Systems, Inc. (SPSi) with support for development and field testing from a consortium effort led by the Civil Engineering Research Foundation, provides the capability to merge the design and construction processes. Using Odyssey, interactive design files can be used to transmit to the constructor the design intent and enable precise capture of these intentions as well as immediate confirmation of the "as-built" conditions.[12]

Besides resolving the frequent discrepancy between what the designer initially draws and what is in fact built, Odyssey and other interactive design-and-construction systems provide substantially improved and very accessible data to the personnel who will later operate and maintain the facility.

CAD technology has penetrated manufacturing and supply segments of the design and construction industry. Custom-designed wooden stairs, for example, are designed on the CAD system and the design information is then fed to automated machinery that manufactures the stair components. Andersen Windows, the world's largest maker of wooden windows and patio doors, places computers and custom software in retailers' and distributors' showrooms, permitting customers to create designs and then develop orders for standard or custom-manufactured Andersen products. The technology has helped Andersen to expand its standard product line from 10,000 models in 1981 to more than 50,000 today.

As helpful as they have been to many designers and constructors, these new CAD and GIS systems are creating what some observers have termed "cultural chaos." Traditional ways of conducting business among owners, designers, and constructors are changing. Contractors are acquiring design capability, and designers, having already input data for their own purposes, find it easier to make quantity estimates and do other tasks normally left to the construction contractor. Issues of potential liability and copyright ownership are arising and may change the ways the industry delivers its products. We will return to those problems later.

The ongoing development of computer-aided design systems illustrates the combined influence of technology's push and demand's pull in producing innovation. Designers could appreciate the potential value of faster computation and more effective data management, and so were immediately interested in the opportunities that digital computers offered. Growing usage of CAD hardware and software created a market that in turn spurred development of increasingly sophisticated and lower cost systems. These new systems expanded the market, and the innovation process continues.

Innovations in Commercial Structures	
Tensile fabric structures	Staggered truss system
Sliding Teflon bearings	Pre-engineered structural
Seismic base isolation	systems
Slurry-wall construction	Tuned-mass damper for
Up-down construction	high-rise buildings
Fall protection on	(drift)
building construction	Active drift control
Composite steel-concrete	systems for high-rise
floor construction	buildings
Metal floor and roof	Blast-resistant (window)
decks	construction
Electrified floor	Anti-terrorist design and
construction	construction
Lateral framing systems	Single-ply membrane
for high-rise buildings	roofing
Precast concrete	Curtain wall construction
construction	Critical path method of
Tilt-up construction	scheduling
Pumped concrete	Ultimate strength design
High- and superhigh-	of concrete
strength concrete	Plastic design in steel
Concrete admixtures	Limit state design in
Concrete floor/deck	timber
hardeners	Sprayed-on fire proofing
Epoxy-coated concrete	Weathering steel
reinforcing bars	Fire retardant ply-wood
Cathodic protection of	Welded-frame system
rebars	scaffolding
Prestressed concrete	Motorized self-climbing
Lift-slab building	scaffolding
construction	Flying formwork
	Gang-forms
	Computer-aided design
	Computer-aided drafting

Source: National Research Council: *The Role of Government Agencies in Fostering New Technology and Innovation in Building*. Washington, DC: National Academy Press, 1992.

BUILDING STRUCTURES

Now consider the design and construction of building structures, another area in which the design and construction industry has experienced significant innovation. Buildings of one type or another predominate the work of the professionals in the design and construction industry, and typically account for more than two-thirds of the total annual volume of U.S. construction, measured in dollar terms. It is a segment of the design and construction industry where many small innovations cumulatively have had substantial impact.

Home-building alone generally accounts for one-third to 40 percent of the volume of U.S. construction. While the structures and other systems in housing (e.g., heating and air conditioning, electrical systems, plumbing) are typically less sophisticated than those of other segments of the industry, one study lists more than 100 specific innovations that have been introduced into residential construction since 1945.

Commercial and institutional structures—e.g., office buildings, hospitals, hotels, and the like—account for a smaller share of the design and construction market, typically less than 25 percent in annual dollar volume, but they often incorporate more sophisticated technology. A less formal survey of innovations introduced over the past several decades in design and construction of such buildings yields a similarly long list of examples.

Of course, not all new technology turns into successful innovation. Efforts to develop single-width brick masonry cladding for high-rise buildings (the Sarabond system) failed when durability proved to be low and parts cracked away from facades.

Sample of Innovations in Permanent Residential Structures, 1945 to 1990	
Functional Area	**No. of Innovations**
Structural exterior wall framing	07
Enclosure and insulation	08
Openings	13
Interior wall framing	07
Foundation	12
Floor framing	10
Roof framing	07
Roof covering	07
Plumbing	12
Electrical wiring	04
Heating/ventilation/air conditioning	12
Interior finish	18
TOTAL	117

Source: Slaughter, 1991.

In comparison, the development of tension fabric structures is an especially interesting case of slow, but successful, innovation. While early applications can be traced to the work of architect Frei Otto in the 1960s, the idea did not achieve widespread or notable commercial use until nearly two decades later, when the advent of Teflon-coated fabrics promised longer life and better resistance to the elements.

The new fabrics can provide lightweight and relatively inexpensive cover for sports stadiums, performing arts arenas, and other large open spaces. Close collaboration among architects, structural engineers, and product manufacturers produced such notable designs as the Hajj airport terminal in Jedda, Saudi Arabia, and the passenger terminal at Denver's international airport.

Lightweight steel stud framing systems, another commercial innovation that has achieved widespread use, were also enabled by a combination of factors. Safety concerns drove the search for a fire-resistant substitute for wood-based framing products, mainly for light commercial applications. A general degradation in the quality and availability of dimensioned-framing lumber, related to the increasing scarcity and rising costs of old-growth timber, lent encouragement

Denver airport. *Courtesy of the S.A. Miro, Inc.; Richard Weingardt Consultants; Severud Associates.*

to this search. The U.S. steel industry was seeking to expand their markets from automotive applications into the building industry. The U.S. gypsum industry was quite willing to cooperate in the design and engineering of a system that would expand their markets for wallboard.

The framing system that was developed revolutionized the management of commercial space. The labor and costs of developing customized building interior layouts to suit an occupant dropped sharply. Interiors are now routinely renovated to fit the wishes of new tenants or the needs of corporate reorganizations. The "service life" of an interior, the period before it is torn out and rebuilt, is now often less than five years. In the last several years, the market forces that spurred the new system's development are encouraging its adaptation to residential housing.

In the case of both tension fabric structures and light-weight steel-stud framing, a change in technology made an older idea attractive in a new application. In the former case, it was the development of Teflon coatings that gave the fabrics greater durability, while in the latter case a market shift brought manufacturers seeking a use for excess capacity together with users who needed a new solution to their problems. In both cases, a combination of circumstances, rather than any single occurrence or condition, was responsible for enabling the innovation to progress. This need for a combination of factors is fairly typical. For example, the automobile would likely not have developed so rapidly had not pavement technology been available to provide appropriate riding surfaces. In turn, there is little question that modern highway pavements have been developed as a result of the demands presented by today's traffic of fast-moving heavy trucks and automobiles. To use our puzzle metaphor, the would-be innovator is likely to be successful in fitting in his or her new idea only if the several surrounding pieces are in place.

SEISMIC RETROFIT TECHNOLOGY

Following both of the major earthquakes that struck California in recent years, news reports have focused national attention on the damage and failures—buckled pavements and cracked buildings—and given scant notice to those structures that did not suffer significant damage. But in fact, there have been significant successes in improving the resistance of buildings and infrastructures to earthquakes. Following the 1971 San Fernando earthquake, the California Department of Transportation began to revise its code requirements for the state's bridges, to increase the ability of bridge columns to withstand earthquake shaking. Research has shown that columns able to deform or deflect in a more ductile manner minimize the risk of extensive brittle fractures that can bring the structure down.

A failed column of the Nimitz Freeway in California following the Loma Prieta earthquake. *Courtesy of the National Institute of Standards and Technology.*

The Robo-Wrapper™ retrofits a traditional concrete column using carbon composites. *Courtesy of XXsys Technologies, Inc.*

As a result, California bridges built to the innovative codes generally performed well in the two recent earthquakes. Even so, the more than 13,000 bridges in California's system include many that are vulnerable, as we saw in the Northridge aftermath. Moreover, the 1989 Loma Prieta earthquake, which struck as the World Series baseball game was being played in nearby San Francisco, collapsed sections of the San Francisco-Oakland Bay Bridge and the elevated structure on the Nimitz Freeway in Oakland. Research has led to development of techniques such as column "wrapping," which involves putting jackets of steel or high-strength fibers—glass and graphite have been used—around a damaged or susceptible column. The jacket provides support similar to that sought by an athlete who wraps tape around a wrist or ankle, and increases the structure's ability to withstand an earthquake. The cost is a fraction of what would be incurred if the structure had to be replaced.

Use of such techniques is termed *retrofitting*, and research has produced another method that has also been adopted for new

buildings. Seismic isolation technologies attempt to reduce or elimi-
nate the sideways shear forces that an earthquake imposes on a
structure by effectively severing the connection between the structure and its foundations. Sliding pads, springs, or other devices, that
support the vertical weight of the structure but absorb a side motion,
are inserted at the base of the structure's columns. Other equipment
is installed as well, some of it quite sophisticated, to assure that the
structure does not slide off its foundations or turn over in a strong
earthquake.

The Pacific Rim Region of the General Services Administration (GSA), for example, is retrofitting the historic U.S. Court of
Appeals building in San Francisco with seismic isolation devices.
These devices, inserted between the building's foundation and base-
ment, rely on a pendulum motion and friction to stabilize a structure
during an earthquake. The building should now be able to withstand
the force of an earthquake measuring 8.0 on the Richter scale, without suffering additional damage to the fixed base structure.

Another example is the 4.3-mile segment of the Foothill
Transportation Corridor in Orange County, CA, opened in April,
1995, utilizing for the first time ever a dual-level seismic-resistant
bridge design. The toll road features an area designed to withstand
two earthquakes of different intensities, a "lower-level" quake and
a "catastrophic" quake.[13] On the East Coast, where residents seldom

U.S. Court of Appeals Building in San Francisco retrofitted with seismic
isolation devices. *Courtesty of SOM, San Francisco.*

realize that they are also exposed to a very distinct earthquake hazard, consideration is being given to retrofitting New York's Queensboro, Whitestone, and Tappan Zee Bridges, and the Delaware Memorial Bridge, using a similar dual-level design approach.

Foothill Transportation Corridor toll road fitted with dual-level seismic isolation devices. *Courtesy of the Transportation Corridor Agencies.*

These seismic designs and retrofit strategies are a result of a focused research effort to solve a recognized problem. The death and destruction caused by the occurrence of major earthquakes near urban areas has been unfortunate, but has probably prompted public recognition of the need for solutions and the value of research to find those solutions. There seems to be nothing quite so effective as a crisis, at least in our present U.S. society, to focus people on the value of doing a better job.

LOOKING TO THE FUTURE

These cases—computer-aided design, building structures, and seismic mitigation technology—are just three of many areas of design and construction innovation we could have discussed. For example, highway safety is a generally recognized need, and improvements have been steady. In the two decades from 1970 to 1990, U.S. highway fatalities decreased more than 15 percent, despite tremendous increases in the number of motor vehicles registered and the vehicle-miles of travel on the nation's roads. The improvements result from a combined effect of many factors, including innovations in highway design geometry, new skid-resistant pavement surface treatments, break-away structural supports for roadside signs, guard rails and median dividers which redirect vehicles onto the roadway and help stabilize their course, crash barriers and exit dividers that absorb the energy of an out-of-control vehicle, and others.[14] During the same two decades, energy efficiency in building heating and cooling has improved with development of more effective insulation materials, new window designs, and better controls on mechanical equipment.

Our point here is that innovation can and has occurred in design and construction. Innovation has brought real benefits that extend well beyond the industry. We assert, nevertheless, that much more can be accomplished. The industry needs to be more aggressive in its pursuit of innovation. It is time to push ahead.

ENDNOTES

1. The following studies seem to us representative of the literature, which we would say is clearly dated: (a) Colcan, M. and R. Newcomb, *Stabilizing Construction*, New York: McGraw-Hill, 1952. (b) Cassimatis, P. J. *Economics of the Construction Industry*, New York: National Industrial Conference Board, 1969. (c) Moavenzadeh, F. and F. Hagopian, *Construction and Building Materials Industries In Developing Countries*, Cambridge, MA: Technology Adaptation Program, Massachusetts Institute of Technology, 1983.

2. Cassimatis, *Economics of the Construction Industry*, p. 2.

3. Subcommittee on Construction Building Civilian Industrial Technology Committee National Science and Technology Council. *Rationale and Preliminary Plan for Federal Research for Construction and Building, November 1994.*

4. U.S. Congress, Office of Technology Assessment, *Technology and the Economic Transition, Choices for the Future, OTATET-283, 1987, Table 4.4, p. 156.*

5. Dean Baker and Todd Schafer. *The Case for Public Investment*. Washington, DC: Economic Policy Institute, 1995.

6. Civil Engineering Research Foundation. *A Nationwide Survey of Civil Engineering-Related R&D*. Washington, DC: Civil Engineering Research Foundation, December 1993, pp. 4-6.

7. U.S. Department of Commerce, Bureau of Economic Analysis, *U.S. Direct Investment Abroad, 1989 Benchmark Survey, Final Results* (October 1992), quoted in Gerald R. Moody, *International Construction: U.S. Industry Condition, Office of Business and Industrial Analysis*, Economics and Statistics Administration, April 12, 1995.

8. National Research Council. *Building for Tomorrow: Global Enterprise and the U.S. Construction Industry*, 1988, and *The Role of Public Agencies in Fostering New Technology and Innovation in Building*, 1992. Both reports cited other studies as well.

9. Based on data generated by Dun & Bradstreet information services, April 1994.

10. Joe Morgenstern. "The Fifty-Nine-Story Crisis." *The New Yorker*, May 29, 1995, pp. 45-53.

11. We use "CAD" to refer generically to a number of systems that sometimes have differing acronyms; e.g., computer-aided design and drafting (CADD), computer-aided engineering (CAE), computer-aided manufacturing (CAM).

12. *Crucial Links for Construction Site Productivity: Real-Time Construction Layout and As-Built Plans*. Final Report prepared for the U.S. Army Corps of Engineers Construction Productivity Advancement Research (CPAR) Program.

13. David B. Rosenbaum. "Design Offers Two Protection Levels." *ENR*, April 17, 1995, pp. 10-11.

14. Imposing lower speed limits, generally found by research and experience to be a highly effective means of reducing fatalities and serious injury, remains controversial.

Chapter 4

PUSHING AHEAD

"Well, in our country," said Alice, still panting a little, "you'd
generally get to somewhere else—if you ran very fast for
a long time as we've been doing."
"A slow sort of country!" said the Queen. "Now, here, you
see, it takes all the running you can do, to keep in the
same place. If you want to get somewhere else, you must
run at least twice as fast as that."

—*Lewis Carroll, <u>Through the Looking-Glass</u>*

The many design and construction innovations that have become standard practice have provided new opportunities to improve our lives, albeit not always without costs. No sooner do we begin to enjoy a new technological marvel, than we find it poses a risk or negative impact, or that it gives rise to the feeling that we need some other previously unimagined product or process. So we face more questions, and need more research. Like Alice, we must keep running, twice as fast as before, if we want to keep up. Unfortunately, there are hurdles to overcome as well.

The race is especially important to anyone in business. There is just no chance to take a break. As soon as you have found a new way to do things, something better comes along. If you do not seize the opportunity, your competitor will. Too many barriers, failures, or hesitations to try something new, and a nation's industry can fall behind its global competitors. Many people feel that the U.S. design and construction industry is falling behind in this way, and our own observations in an international setting have convinced us there is some cause for concern. But before we try to deal with that issue, consider the opportunities we have before us. They are another element of the innovation puzzle.

We do not need great prophetic powers to recognize that in the future, designers, craftsmen, laborers, and managers will have to bring new skills to construction. Future plumbers and electricians are likely to have to deal with biological processes and electronic components that formerly were the concern of chemical and electrical engineers. The basic work of carpenters and welders will change as new structural materials are introduced. The construction process will continue to evolve as new mechanical and electronic devices increase the accuracy and productivity of construction. Lasers and digital measuring devices, already common on many construction sites, will become universal. Combined with electronic equipment controls, future excavators, cranes, and other construction equipment will become "smarter," safer, and more accurate.

Worldwide economic and political changes form a backdrop for such developments. People are migrating from place to place and business conditions are changing quickly, requiring greater flexibility to shift resources to where they are needed. Such increasingly popular terms as "adaptive reuse" and "retrofit" signify the creative redeployment of built assets that these trends necessitate. Companies and nations that once were intense rivals find themselves doing business together. They share risks by seeking ways to work efficiently together. Combined with movements of skilled labor in pursuit of jobs, the design and construction industry will undoubtedly continue its evolution toward multi-cultural, high-production activities.

In our visits to research labs and sites of innovation in the United States and abroad, we have seen that some of these trends are far progressed. It does not take a wild imagination to recognize that developments such as the examples in the following pages will have important influence on the industry.

HIGH-PERFORMANCE MATERIALS

While the research on concrete and other construction materials has not yet moved into the molecular level, work is under way to improve materials performance. One major effort to promote advances in construction-oriented material science is the high-performance CONstruction MATerials and Systems program, known as CONMAT. The Civil Engineering Research Foundation (CERF) brought together ten industry sub-groups (representing aluminum, steel, wood, hot mix asphalt, masonry, concrete, fiber-reinforced polymer composites, roofing materials, coatings, and "smart" material devices and monitoring systems) with the goal of stimulating efforts to create the materials and systems for a new generation of constructed facilities. These separate industry sub-groups seldom work together, but in the CONMAT program they have laid out a two-billion dollar program of research, development, and deployment to be carried out over the next ten years.[1] While all of these material industry sub-groups have identified a whole host of exciting projects and concepts that are either underway or under consideration, space permits our mentioning only a few of them below.

The Paris Landing Bridge in Tennessee uses high-performance steel that reduced the structure's weight and saved 32 percent of the estimated cost of a conventional span.
Courtesy of U.S. Steel.

For example, advanced process control technology and microalloying (which involves very fine control of alloy composition, almost to the molecular level) can in theory create steels with improved ductility and toughness that would substantially improve the safety of buildings and bridges likely to be exposed to seismic forces or other situations where higher energy absorption capability is wanted. These methods may also reduce costs to a level that stainless steels could be used to extend the lives of structures such as highway bridges and water treatment facilities that must withstand particularly corrosive environments.

Aluminum alloys offer the advantages of light weight, ease of fabrication, and excellent corrosion resistance. While these advantages have motivated the use of aluminum in military applications, they have generally been outweighed by high initial cost for civilian highway bridges. New production methods and greater attention to life-cycle costs may change that.[2] In short-span bridges, for example, aluminum's light weight and low maintenance costs could make the material an economical choice for bridge deck replacements.

Aluminum roof for waste water treatment clarifier. *Courtesy of The Aluminum Association; Conservatek, Inc.*

For longer spans, erection times that can be as little as one-third those for conventional bridge systems and increased load-carrying capacity may be decisive in selecting retrofit strategies. Removal of weight from the suspended deck (i.e., through use of light-weight aluminum, composites, or other light-weight materials) could actually permit an increase in safe traffic loads on old bridges as well as reducing seismic damages by reducing inertia forces. In addition, the ability to preassemble large sections of the light-weight deck and move them into place should substantially shorten total construction time, with an associated substantial cost saving. Research on metallurgy, fabrication, and design will be needed to accomplish these advances.

Fiber-reinforced polymer composites, with a solid record of successful application in space-flight vehicles and missiles, military and commercial aircraft, and automotive components, have enormous potential for use in construction applications. Their combination of high strength and light weight, design flexibility, and corrosion resistance characteristics make these composites particularly appealing for such applications as replacement bridges for highways and other transportation applications. Researchers envision that such bridges could be fabricated in factories, with exceptional quality control and costs comparable with traditional precast and pre-stressed concrete designs, and transported over-the-road to be opened to traffic in one eight-hour working day. Drivers on heavily travelled roads will ap-

Devil's Pool in Fairmount Park, PA is home to the longest fiberglass pedestrian bridge (50' 0") in North America and the third longest in the world. *Courtesy of Creative Pultrusions, Inc.*

preciate the benefits! Some companies are already demonstrating this concept on short-span replacement bridges for local country-road applications. Creative Pultrusions, Inc. and the Maunsell Group, for example, have demonstrated the concept for industrial customers with construction of bridges at least 30 feet in span-length, and much longer-span bridges are being planned.

Another potential application of composites is electrical transmission and distribution structures (e.g., transmission towers, utility poles). Such structures are non-conductive and resist corrosion in seaside installation, major advantages for situations where high-voltage electricity is present. A demonstration has shown that an eighty-foot composite transmission tower can be erected by a two-person crew in one day, far less labor than the typical five-day schedule that a traditional approach entails.

Polymers and resins, the matrix for many composites and, by themselves, widely used in electronics, consumer products, automotive, and other smaller scale applications, may be considered "high-performance" materials for design and construction, especially large public works infrastructure. Use of high- and medium-density polyethylene (HDPE and MDPE) and polyvinylchloride (PVC) in underground piping networks has been growing rapidly, because of these materials' corrosion-resistance, flexibility, and declining cost relative to competing materials. The majority of new natural gas distribution piping put in place is now made of PVC. Liquid polymer and membranes are being developed and used to stop leaks and seal cracks in older underground lines. Such materials will save millions of dollars in maintenance and repair efforts and reduce disruptions of traffic and businesses. Similar materials will increasingly be used to seal waste repositories and prevent infiltration of corrosive salts on bridge decks and other transportation structures. Corrosion loss in the U.S. amounts to $250 billion annually.

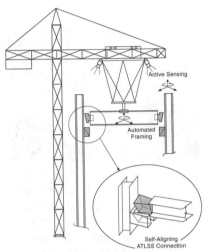

The ATLSS Connector aides in the construction process by aligning beams to columns using automated construction methods and robotic cranes. *Courtesy of Center for Advanced Technology for Large Structural Systems (ATLSS), Lehigh University.*

AUTOMATION AND INSTRUMENTATION

As in other sectors, microcomputers and other electronic instrumentation and controls are finding increasing application in design and construction. Advanced sensing and control devices will be used in the harsh environments of construction sites to improve construction accuracy and workers' safety. Remote and robotic control of heavy construction equipment will facilitate maintenance and repair in dangerous and hard-to-reach situations. We have described some of the innovative technology already being demonstrated, but we anticipate much more dramatic developments.

For instance, the Center for Advanced Technology for Large Structural Systems (ATLSS) at Lehigh University, in cooperation with the National Institute of Standards and Technology (NIST), has recently completed a 5-year comprehensive automated construction research program for an integrated building system. The program focused on a new self-aligning beam-to-column connector for use in automated construction and robotic cranes capable of improved placement of members using the new connector. The ATLSS Connector (AC) is expected to minimize human assistance during construction and provide a quicker, safer, and less expensive erection procedure.[3]

New developments will extend to facility operation as well. Systems control and data acquisition (SCADA) systems, already being widely adopted for municipal water supply and natural-gas distribution system management, will become universal in urban and regional infrastructure. Similar technology—e.g., remote sensors, telemetry for data transmission, and computer-based simulation and decision-support algorithms—will enhance energy efficiency, occupant safety and comfort, and security in building complexes.

Miniaturization of mechanical devices seems likely to enhance both efficiency and comfort in building climate control. Tiny thermometers, heat pumps, and blower motors produced by "micromanufacturing" processes now being developed in laboratories could replace large air-conditioning units with banks of wall-mounted chips.

Annually, fires—in homes, other buildings, and infrastructures—cause deaths and injuries, millions of dollars in damage, and disruptions in economic activity. While prevention will continue to be the primary focus of efforts to reduce losses, advanced sensing and monitoring systems, combined with simulation, will facilitate earlier detection and faster and more effective response when fires do occur.

These various applications will produce large quantities of data that can be used to develop better mathematical models to predict facilities' lifetime performance and specific response to use. Such models will support investment and asset-management decisions and monitoring of damage in structures, from such causes as seismic events or normal use and wear.

"GREEN" TECHNOLOGIES

Increased environmental awareness and the search for "sustainable" living patterns have produced a great deal of discussion of "green" technologies. Although some people use the term to refer to low impact on the natural environment, while others may use it for recyclable materials, there are no generally accepted definitions or measures of "greenness." It clearly includes considerations such as the use of less energy, the reliance on renewable resources, the preservation of wildlife habitat, and the broad compatibility with the "natural" environment.

Solar panels are just one of the methods being explored to provide a sustainable source of energy. *Courtesy of U.S. Department of Energy.*

In design and construction, the concept has been reflected in uses of living plants and terrarium-housed plant and animal communities to condition indoor air for offices, and "cradle-to-grave" analyses of the environmental and energy-use consequences of using popular building materials. Development and use of biodegradable solvents, paints, and lubricants, and increased efforts to control dust and sediment in stormwater and other runoff from construction sites are examples of "green" technology sought at a smaller scale.

The concept will undoubtedly continue to provide opportunities for innovation. Design and construction technologies likely to emerge include more effective recycling or re-use of waste and worn-out materials (e.g., highway pavements and demolition debris), more energy-efficient construction methods, advanced pollution prevention and control methods, and a progressive shifting toward wood and other materials that are presumably renewable.[4] In addition to specific processes and products used in design and construction, we anticipate that the basic values and assumptions underlying how the industry operates will have to change in some important ways.

The principles of life-cycle design, for example, now only used on a limited basis to refer to the period from a facility's design to the end of its economic life, will have to be broadened. Principal resources used in a facility will be tracked from mining or manufacturing through cycles of reuse to ultimate disposal. Minimizing abandonment or deposition in landfills or offshore dumps will become a primary measure of design and construction performance.

Land must come to be viewed as a resource to be managed in the same manner as other limited resources. While the question of whether land as a resource is renewable or non-renewable is still open to discussion, new techniques are being devised for "low impact" development, designed to work with, rather than overwhelm, natural ecosystems.[5] Such techniques, which may be as simple as not rebuilding in floodplain locations following a disaster, maintaining natural water-flow patterns and runoff volumes, and conserving critical wildlife habitats, will become standard practice. More complex techniques may involve management of urban subareas as "closed systems," with district energy management and complete on-site recycling of water and wastes.[6] Demonstrations in Europe, for example, suggest that the practice could extend to complete "industrial ecosystems" in which the treated wastes and their byproducts can be used as the inputs from one industry to another.

Difficult questions have yet to be resolved in this search for "green" technologies. Wood, for example, while still the primary building material for residential structures, is increasingly in short supply and often of poor quality. Traditional residential building practices prefer slow-growing tree species and "old growth" timber because of the greater strength, dimensional stability, and yield of finished lumber derived from such sources. Environmental restrictions and depletion of forest reserves have made lumber prices one of the principal contributors to rapidly increasing costs of residential construction.

High-performance engineered wood products are one response to these developments. (Another is the shift to steel-stud

Wooden three span "tridge."
Courtesy of the American Institute of Timber Construction.

framing we previously cited.) Manufactured I-joists and box beams that use different types of wood in the structural section to achieve optimum performance and laminated wood (e.g., plywood and laminated beams) are now increasingly used in non-residential construction as well. Wood is also being used in conjunction with polymer- and resin-matrix composites, for example, in fiber-reinforced glued laminated beams, and wood trusses that incorporate metal webs. Such developments are likely to continue, enabling more effective use of plantation-grown wood.

CLEANING UP AND KEEPING UP

In contrast to the search for "green" technologies, past practices of development and industrial activity have left us with an extraordinary number of sites where hazardous and toxic materials have been dumped and abandoned. Increased awareness, public concern, and a growing body of legislation and case law severely restrict reuse of these sites. Accumulated experience and innovative technologies are helping to clean up or contain the wastes and make the sites safe:

● Geotechnical barriers that have high strength, very low permeability, and high durability are being improved to contain wastes and sharply reduce the risk of contaminating adjacent water supplies.

● In-situ vitrification is being perfected, using electric current to melt contaminated soils into blocks of glass-like material which safely contains the pollutants.

● Organic stabilization shows promise, mixing micro-organisms into contaminated soils to convert wastes to more benign forms .

● Groundwater modeling, the application of high-performance computing to model the migration of pollutants in groundwater, will enable more informed decision-making about the risks at particular sites and the relative merits of alternative remediation strategies.

Whether we are talking about new technologies to clean up sites and reduce hazards to our health or programs designed to improve public safety, the business opportunities for the entrepreneurial individual or company, as we move into the twenty-first century, are staggering. For example,

Target Benefits of Intelligent Transportation Systems (ITS)

- Better travel information
- Quicker emergency response
- Easier travel
- Improved traffic flow
- Fewer traffic jams
- Improved trucking management
- Faster freight deliveries
- Reduced pollution

Source: ITS America.

safety is one of the key concerns motivating programs like the Intelligent Transportation Systems (ITS) in the United States. ITS seeks to utilize advanced electronics and computer technology to develop highway systems that integrate "smart" cars and highways for safer, more efficient driving. The concept is intended to equip your car with features such as a warning system which sounds an alert when your car gets too close to another or an electronic accident system, which would direct emergency personnel to the accident site if an accident were to occur. Proponents of this intelligent transportation system estimate that within 15 years it could save at least 3,300 lives and prevent 400,000 injuries annually. Investments in new businesses around the country to develop technologies to support "smart" car and highway systems is becoming a major growth area in the U.S. The benefits of an innovative system such as ITS will affect us all, and will bring the transportation system into the information age, thereby improving productivity, efficiency, and quality of life.

The opportunities and needs for design and construction innovation are here, and we think they are growing. We have been successful innovating in the past, but have not been making the effort that seems warranted to increase innovation now.

What makes us especially sure that U.S. industry should be doing more is what we have seen in other countries. Our visits to Japan, Europe, and elsewhere make it plain that sustaining innovation has global implications. Huge markets await successful design and construction innovators. What we have seen overseas also offers some lessons in how to piece together the innovation puzzle.

ENDNOTES

1. Civil Engineering Research Foundation. *Materials for Tomorrow's Infrastructure: A Ten-Year Plan for Deploying High-Performance Construction Materials and Systems Executive and Technical Reports*. Washington, DC: Civil Engineering Research Foundation, 1994.

2. Designers would prefer typically to think in terms of "lifecycle" costs, which include operations and maintenance along with initial construction costs when decisions are made. First costs, however, often are limited by budgets and political forces.

3. R.B. Fleischman, Le-Wu Lu, B.V. Viscomi, and K. Goodwin. "Design and Implementation of ATLSS Connections." Presented at the First National Conference and Workshop on Research Into Practice, June 15-16, 1995, sponsored by the National Science Foundation and the University of Maryland.

4. Some people question whether even wood is truly renewable under current production methods. They note that "old-growth" timber is not being replaced and that the quality (e.g., density and stability) of faster-growing plantation-bred trees is not equivalent. Some people also argue that the cycles of cutting and replanting entail soil erosion, chemical pollution (with fertilizers and pesticides), and loss of wildlife habitat that fail the test of sustainability. Others disagree.

5. The debate often rests on whether land can be fully returned to a "pristine" condition, within the context of human use. Within limits, land can certainly be recycled, as the real-estate industry well knows.

6. District energy management involves common supplies of energy and the recovery of waste heat within defined geographic areas, e.g., an urban downtown. The practices are perhaps most widely used in Scandinavian cities.

Chapter 5

INTERNATIONAL EXPERIENCE

Over the past several years, we have had opportunities to see how the design and construction industry in other countries conducts its business. Our professional positions have given us access to the experience of many of our colleagues in other countries, and to the observations of our compatriots who travel as well. We have travelled to Asia and Europe to see and hear first-hand about emerging technologies and how they move into practice. We have visited less developed countries where the differences between people's aspirations and the current quality of life are enormous.

We have found that these experiences offer valuable insights for how innovation can occur. More importantly, from our perspective, the practices and institutional relationships that work in other places offer valuable lessons for how we can do a better job of introducing innovation in the U.S., by targeting real opportunities and overcoming impediments to design and construction innovation. The construction industries and governments in other countries have developed policies and mechanisms for transferring new ideas from concept and laboratory into practice that provide models for what we can do in the United States.

Our experience has helped us learn about what is being done, how different countries approach the marketplace, and their underlying business philosophies. Each country operates differently, in part because of differences in materials, climate, and geography, but also because of differing social structures, history, and culture. Each country has its own strengths and weaknesses. Investments in innovation are likely to yield different rates of return at different times and in different regions of the globe. Yet there are commonalities as well and there are lessons to be learned. Let us start by reviewing some of what we have seen.

ON THE COMPETITIVE EDGE

There has been much discussion over the past years, in professional circles and the press, about the U.S. design and construction  industry's performance in international competition. Following the Second World War, U.S. industry led the rebuilding of wartorn nations and then profited handsomely as the world's attention turned toward the needs of economically less developed countries for infrastructure. While design and construction never accounted for a major share of the U.S. export trade, hundreds of companies followed behind the bulldozer to invest in capturing the new markets that were opened up. At home, design and construction were solidly all-American.

Things changed in the 1970s and 1980s. The U.S. share of the international design and construction market, measured in dollar-value of contracts, declined and annual U.S. exports of construction equipment fell by two-thirds from its 1978 peak of $6.3 billion. Employment in that segment of the industry followed suit.[1]

At the same time, foreign competition entered the U.S. domestic design and construction market. Firms from Western European countries began to work in the United States and to invest in U.S. design and construction companies. Japan's large construction companies appeared on the scene as well, as Japanese investment in U.S. real estate increased and Japanese manufacturers began to produce goods here.

As we and other U.S. industry observers studied the situation, we found that many of these foreign firms were outspending their U.S. counterparts for research and development, both in terms of absolute expenditures, and relative to annual sales and employment. More importantly, the spending seemed to be paying off in the high quality of the finished facilities and in the technology employed. We and others began to express concern that U.S. industry was falling behind in design and construction, just as it had in the automotive and consumer electronics industries.[2]

In response to mounting concerns, U.S. industry groups and government agencies took a closer look. Trips planned and sponsored by the National Science Foundation, the Department of Transportation, the Civil Engineering Research Foundation, and others sent U.S. professionals to observe first hand how the construction industry operates in those countries and where they stand regarding development of new technology.[3] These groups found that in some areas of design and construction technology, U.S. industry is the leader, but in other areas, that leadership has been assumed by others (see table).

U.S. Leads	Europe Leads	Japan Leads
High-performance concrete	High-performance asphalt	High-performance steel
Waste/wastewater treatment	High-speed rail/Magnetic levitation vehicles (Mag-lev)	Automated equipment
Computer Aided Design (CAD)/ Computer Aided Engineering (CAE)	Tunneling	Field computer use
Solid/hazardous waste disposal	Real-time site positioning systems	High-speed pavement assessment
Environment	Restoration	Safety
Geographic Information and Positioning Systems (GIS/GPS)	Marine construction	Intelligent Buildings
Integrated databases	Energy conservation	Building systems

Relative levels of technology adoption in design and construction industries, as assessed in reconnaissance trips sponsored by the U.S. National Science Foundation and the Civil Engineering Research Foundation. Source: Civil Engineering Research Foundation European Research Task Force. *Constructed Civil Infrastructure Systems R&D: A European Perspective.* Washington, DC: Civil Engineering Research Foundation, 1994, p. 5.

These groups also found, however, that in a number of those areas where the U.S. industry still leads, the competition is catching up.

In addition, these groups found that the ways the design and construction industry operate in some of these countries offer a refreshing change from the problems present in U.S. practice. Unified national markets operated with a single set of codes and regulations protecting public health and safety, compared to the patchwork of local- and state-government controls in the United States. Designers, constructors, and their clients seemed more frequently to work together toward common goals, compared to the inherently adversarial relationships that U.S. bidding and contracting practices have spawned. The industry and its participants also seemed to enjoy greater respect and to have more significant roles in setting their nation's policies on investment in infrastructure and housing, export policy, and other areas important to the industry. Was there some way that U.S. practices could be shifted to be more like that?

BUILDING CONSENSUS IN JAPAN

Like other U.S. visitors to Japan, we noted that that nation's corporate philosophy and government policies stress cooperation and the development of consensus as a basis for action. For the Japanese design and construction industry, that means bringing together contractors and consultants, product and material suppliers, government, and academia to work toward common goals. This coordinated approach promotes useful research and helps translate research results into practice.

The large highly integrated firms that characterize Japan's construction industry, which stands in sharp contrast to that of the United States, facilitate innovation. The Japanese government has a strong influence on major research and technology application initiatives and maintains a methods- and products-approval process that supports domestic innovation and acts effectively to discourage foreign competitors. In fact, Japan's Ministry of International Trade and

Industry is planning on doubling the country's research and development spending by the year 2000. This means Japan's national R&D budget would grow from approximately 5 percent a year to at least 10 percent.[4] We think it is important to note, however, that the Japanese construction firms also play a central role in direct funding of research. For instance, in 1992 the five largest Japanese construction companies, defined in terms of total revenues (Kajima, Shimizu, Obayashi, Taisei, and Takenaka), committed research and development funds totalling approximately $700 million, more than double the $343 million the entire U.S. design and construction industry was estimated to spend. While Japan's notoriously high prices may skew such comparisons, the overall scale of design and construction in the two countries is about equal. Tax incentives provide a partial explanation for the Japanese industry's high rate of spending, as indicated by the following, to encourage design and construction research.[5]

- When a Japanese firm acquires, produces, or constructs a plant or equipment that is more energy efficient or more state-of-the-art, and uses it within one business year, the company receives an investment tax credit worth 7% of the unit's cost or 20% of the firm's income tax, whichever is less.

- If a Japanese company spends more on R&D in the current year than it did in the previous one, the firm receives a tax credit equalling 20% of the difference or 10% of the corporation's tax before tax credits (typically 30-40% of taxable income), whichever is less.

- Japanese companies receive tax credits for the development of certain projects: shorter depreciation lives for assets contributing to prevention of disasters caused by earthquakes (15%); qualified high rises in specified city planning zones (24%); and qualified assets for research and development located in designated areas (30%).

● The Japanese government allows for a tax-free reserve for a construction company to provide for additional costs in repairing defective portions of their work (based on actual costs for two preceding years or 0.5% of construction cost).

A significant force in the Japanese R&D process is Japan's Ministry of Construction's long range "visionary" concepts, that stimulate research to meet future national needs. For example, the Japanese have undertaken major research efforts to solve problems posed by land shortages, crowded conditions, frequent seismic events, severe weather, and shortages of skilled domestic labor. There is also a degree of coercion in the sense that the Japanese culture, which differs significantly from that found in Europe and the U.S., requires large contractors to demonstrate competence with the latest building technologies to qualify for large public works contracts. As an additional incentive, Japan's cooperation-based system for managing liability for loss and injury reduces the risks involved with introducing new technologies.

HARMONIZING EUROPEAN DIVERSITY

The independent nations of Western Europe historically represent a diversity of economic and regulatory conditions at least equal to that of the United States. With the formation of the European Union (EU), these nations are moving aggressively to overcome many of their differences and establish a relatively unified market for construction and infrastructure which will ultimately exceed the U.S. market in aggregate size. The lowering of trade barriers and "harmonization" of regulatory controls that this move entails is being accomplished using a variety of mechanisms that are spurring technological innovation. In addition, several of the individual governments have long supported their construction industries' innovation efforts as a key to stronger exports and an instrument of foreign policy. In Sweden, possibly the most extreme case, industry has imposed on itself a three-cent-per-hour payroll tax that is paid into a fund to support research by the industry. Individual companies submit proposals for research funding which is supplied by the approximately $6 million the tax yields annually.

One particularly exciting development is the French "innovation charter" program to protect new technologies long enough for the developer to recoup its R&D investments. Innovative products, processes, and equipment that are not specifically covered by standards or appraisal certificates can be granted a charter by the government that provides for exclusive rights to these new technologies for several years. Innovation charters allow contractors to team up with government research institutes when the contractor does not possess the technical expertise or equipment to develop the innovation alone. To make the innovation charter system work, contractors are encouraged to propose new ideas supported by experimental data. Because more emphasis is given to quality and durability during selection of contractors, what actually makes the system work is the fact that publicly funded projects are not required to take the "low bid" on initial costs, but rather the "best bid" based on life-cycle costs.

In almost all countries in Western Europe, the national governments take a leadership role in offering incentives for industry to work together as a team with universities and government organizations to pursue technological advancements. Multinational research funds have been pooled into major international technology initiatives such as BRITE/ EURAM (for development of advanced materials), THERMIE (for new energy technologies), ESPRIT (for information technology), and EUREKA (targeting general construction competitiveness). These pooled research funds represent an incentive for industry-based research. This system requires companies to work across national borders, teaming with other companies and forming partnerships with academic communities, to have access to the pooled funds.

For the members of the EU, "harmonization" includes transforming a multitude of national standards, now used in construction and manufacturing, into a single coordinated regulatory framework. The EU's Construction Products Directive was issued to assure that national building codes and product standards do not inhibit trade within the Union or the EU's ability to compete effectively in international markets. The European Council for Standardization (CEN) is the primary forum for the harmonization effort. The European Organization for Technical Approvals (EOTA) was created to produce European Technical Approvals (ETAs) for which there exist no European harmonized standards. Representatives from the European Free Trade Association (EFTA) can participate, making the effort essentially all-inclusive for what is likely to become the world's largest single market for design and construction.

Their goal is to devise a single European Standard by 1996. This date may be optimistic, but there is no doubt that the process of inter-country comparisons will winnow out the best from each country's system to create a broadly acceptable set of standards. To accommodate innovative products that fall outside the standard, some European countries have adopted the Agrément system, a product approval process similar to the Japanese system for certifying new technology and facilitating its move into practice. A technical "Agrément" organization provides technical assessment services for clients seeking evaluations and certifications of the new product's acceptability in their specific country.

The European Standard, when developed, is likely to gain wide acceptance outside of Europe as well. Other countries have already begun to adopt product-review systems similar to the Agrément. The Civil Engineering Research Foundation has already identified over 25 countries from North and South America, Europe, Asia, Africa, Australia, and the Pacific Islands which have organizations directly involved in the technical approval and evaluation of new innovative products being introduced into the construction industry marketplace. These technical assessment organizations may utilize approval criteria developed by other organizations, rather than devise their own approval criteria as the European organizations have done. There are significant variations in such areas as scope and authority of the evaluations, processes of evaluation, types of evaluators, reporting mechanisms, and liability concerns. They nevertheless represent a substantial improvement over the tangled net of local regulations that has characterized the industry in many countries, including our own.

PROJECTS AS TESTBEDS

We observed in both Japan and Europe that large and technically challenging projects are presenting unparalleled opportunities for research and testing of new technology. It has become increasingly difficult in the United States to justify and undertake major construction projects of the sort more commonly found overseas. While we are currently hard at work on such mega-projects as Boston's Central Artery, Third Harbor Tunnel, and Deer Island sewage treatment plant, and the new Denver International Airport which has opened for service, the foreign examples are huge and numerous. The Anglo-French Channel Tunnel (popularly called the "Chunnel"), Stockholm's largely subsurface Ring Road, the Tokyo Bay Causeway, Kansei Airport, and very tall buildings being developed in southeast Asia—to give a few examples—currently represent the most challenging work for designers and constructors. Innovation on such projects is demonstrated and given exposure to a broad audience of users and policy-makers as well as industry professionals.

The revolutionary Channel Tunnel, "Chunnel", connects the United Kingdom and France. *Courtesy of Lemley & Associates, Inc.*

The reasons why we have fewer such projects in this country are numerous. Less centralized government controls make approvals harder to obtain. People are inclined to protest any major undertaking. Other reasons could be given as well, but we certainly do not mean to suggest that research opportunities are a pivotal objective for initiating these costly, risky, and often disruptive projects. We do believe that our lack of such opportunities means we must try harder to learn as much as we can from the design and construction that we do undertake.

WORKING TOGETHER

This learning, we observe, is often a cooperative effort by multiple firms, government agencies and laboratories, and academic institutions working together in consortia. The organization and strategic direction of these consortia may be established by industry bodies, as is largely the case in the European Union, or by government, as is typical in Japan. In both regions, there is a strong interaction of government and private expertise and significant cooperation among firms that are otherwise competitors. The combining of resources that occurs is primarily only found in the defense and aerospace industries in this country. The result of this cooperation is that the best people from many institutions have common direction. In addition, the inevitable cost of negative research findings is shared among all participants.

A similar spirit is observable in the use of peer review to assess technological risk in new technology. In both Japan and the European Union, promising new ideas may be submitted to an institution that assembles panels of experts that include representatives of industry, government, and universities. These panels review available information, request that tests be conducted and, when they are satisfied, effectively certify that the balance of risk and reward warrant broader application of the new technology. With this endorsement, facility owners and the public at large more willingly accept that the risk of the new technology is reasonable.

The U.S. design and construction industry seems to find these cooperative approaches difficult, perhaps because of concerns about government's anti-trust rules intended to discourage monopolistic business practices, or because competition is closer to the emotional and philosophical core of U.S. business practices than in most other countries. Perhaps we are simply unable to depart from the concept of individual responsibility and the need for someone to blame when anything goes wrong. The relatively recent emergence of new institutions or programs fostering industry, government, and academic cooperation—e.g., the National Science

Consortia building links talented individuals from many different disciplines and organizations to get a job done.

Foundation Civil Infrastructure Systems Initiative, the Strategic Highway Research Program (SHRP), the Civil Engineering Research Foundation (CERF), the Construction Industry Institute (CII) and others—may signal a change in U.S. attitudes.

On our visits to Japan and Europe, we also observed a greater reliance on expertise and experience of constructors as well as designers in the facility development and management process. Lowest cost is seldom the single primary factor in selection of firms to undertake project development. In some cases, construction companies are invited to qualify for bidding based, in part, on their research and applications of pioneering new technology. As a result, individuals and companies are rewarded for their ability and willingness to achieve innovation. In several European countries, contracts are often awarded on the "best bid" basis rather than least initial cost, i.e., they consider likely overall contractor performance in terms of effectiveness as well as price. Bidders can thus offer innovative technology at higher cost if the value can be demonstrated.

This demonstration of value depends on all parties having a reasonably firm idea of what they mean by "performance." We have observed that European and Japanese owners, designers, and constructors are not necessarily any more certain than we are, in advance of construction, of what they will call "acceptable performance," but they do seem more willing to work together to assure that acceptable performance is achieved. One result has been that performance has developed further as a measurable concept, particularly in Europe, and performance specifications are increasingly being used in construction and rehabilitation. (The Netherlands, since 1992, has had the only performance-based building code in the world, although New Zealand has now adopted one, and other countries are said to be considering the change.) Constructors consequently have the opportunity to propose innovative processes and products that will deliver the required performance, perhaps in return giving the owner added assurance, in the form of extended performance guarantees or warranties. In addition, design and construction are more frequently undertaken within a single contract, even for public infrastructure. As a result, top-quality designers and constructors have more opportunities during the facility-development process to apply innovative technology.

LESSONS FOR CHANGE

A number of common themes have emerged from our overseas observations, related to forces driving innovation in Japan and Western Europe. As the world moves more and more toward a truly global economy, infrastructure and construction are becoming even more competitive global markets. More innovative companies and workers—in Japan, Europe, and elsewhere—are learning new processes and developing new products that will find receptive buyers in the marketplace.

We are personally committed to the ideas that international cooperation is good and that global markets will spur greater innovation in design and construction. We are concerned, nevertheless; if we fail as a nation to demand innovation of our own design and construction industry, and if our design and construction industry fails to deliver, the nation will purchase its new technologies elsewhere or our quality of life will suffer. If we as a nation cannot be equal technological partners, we will lose out. As Benjamin Franklin said regarding his fellow rebellious colonists' approach to protesting British management, "We must hang together, or we shall most assuredly all hang separately."

To some extent, U.S. business practices, tradition, culture, and other characteristics may limit the application of models and experience from other countries' construction industries. We believe, however, that many of the lessons from abroad are transferrable or adaptable.

These lessons can become the basis of strategy for enhancing innovation in U.S. design and construction. The strategy, to be successful, will have to involve all major stakeholders as participants in solving the innovation puzzle; this is the first key lesson overseas experience has

> **Common Themes in Overseas Practices Are Drivers of Innovation**
>
> - Major construction projects as research laboratories and testbeds for new technology
>
> - R&D activity undertaken jointly by multiple firms, government laboratories, and academic institutions, acting within a strategic framework established by industry bodies or government
>
> - Peer review of promising new ideas, by industry, government, and academic experts who reduce perceived risk by effectively certifying that the balance of risk and reward warrant broader application
>
> - Qualifying construction companies based on "experience and expertise" demonstrated in part by research and applications of pioneering new technology
>
> - Using performance specifications, sometimes matched by extended performance guarantees or warranties
>
> - Awarding contracts on a "best bid" basis rather than least initial cost, i.e., considering the likely overall performance as a function of price and effectiveness
>
> - Integrated contracting for design and construction, giving top-quality designers and constructors more opportunities during the facility-development process to apply innovative technology

to offer. A closer relationship among owners, designers, constructors, and the many others involved in facility development, throughout its service lifetime, pays off in greater opportunity to innovate since key decision-makers are there when innovation opportunities arise.

These decision-makers can then act quickly to take advantage of the opportunities. It is, nevertheless, the facility owners, in both the private sector and government, who must be the primary motivators for innovation. Their demands for better performance, in terms of shorter delivery times, lower life-cycle costs, and improved standards of service, are the most critical messages for designers and constructors. These latter participants hold a stake in innovation that will improve their own working methods, but the demands of competition and clients both raise that stake and offer added payoffs for finding ways to improve the product as well as the process. Some U.S. owners are consistently willing to pay for facilities expenditures that may yield a good long-term return, e.g., large corporations or public infrastructure authorities that build for their own use, intending to retain long-term ownership of facilities to be used by others. The cases are, however, sufficiently rare as to be noteworthy. When the insistence on better performance becomes routine, as we believe is more typical in Japan and becoming more so in Europe, designers and constructors will routinely respond.

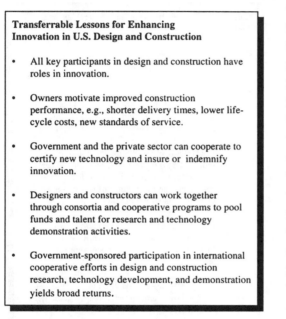

Transferrable Lessons for Enhancing Innovation in U.S. Design and Construction

- All key participants in design and construction have roles in innovation.

- Owners motivate improved construction performance, e.g., shorter delivery times, lower life-cycle costs, new standards of service.

- Government and the private sector can cooperate to certify new technology and insure or indemnify innovation.

- Designers and constructors can work together through consortia and cooperative programs to pool funds and talent for research and technology demonstration activities.

- Government-sponsored participation in international cooperative efforts in design and construction research, technology development, and demonstration yields broad returns.

Regardless of cooperation and strong motivation to innovate, developing and applying new technology is a risky business, and managing that risk is essential. In this country, where litigation is a way of life, risk management is not a job that private business can handle alone. The European and Japanese experiences of joint government and private-sector effort to review new technologies and effectively certify that risks are reasonable within the context of current practice can be supplemented by insurance and indemnification for firms seeking to innovate.

In a setting where there is cooperation and trust among the parties of a project, the practice of constructors effectively "self-insuring" becomes viable. The constructor takes responsibility with the understanding that owners and sometimes government will help out if an unforeseeable problem arises. While this approach is popular overseas, only large firms can withstand substantial problems. In the U.S., formalized insurance distributes some of the risk and enables competent smaller firms to survive what might otherwise be disastrous events and litigation.

To the extent that the cost of constructors' liability insurance can be passed on, facility owners now pay a share. It would be a more effective arrangement, however, if owners willing to sponsor innovation could purchase their "innovation insurance" separately. A pooling of many projects would distribute the risks of each project among many innovators. The costs for coverage, the premium, could be determined by the underwriter's assessment of the relative level of risk inherent in each specific case. Government sponsorship might be required to establish the program (e.g., in a form of guarantees that would be similar to federal insurance of the public's deposits in financial institutions), but the costs would be borne entirely by the design and construction industry and its customers. To keep those costs reasonable, an arbitration system could be incorporated in the insurance program to reduce litigation when problems do occur.

A partial adaptation of this lesson is already found in the recently-formed Highway Innovative Technology Evaluation Center (HITEC) established by the Civil Engineering Research Foundation (CERF). This is the first of several planned initiatives designed to provide a forum for all participants in the design and construction industry to work together to test, demonstrate, evaluate, and document innovative technologies. Established through cooperative agreements between government agencies, industry associations, and private companies, the center uses existing national laboratories, university-based research institutes, and private companies to test, display, and then promote application of new technologies in the highway transportation field.

Programs like HITEC comprise one of the several models that owners, designers, and constructors can emulate through formation of consortia and cooperative programs. These joint activities, modeled to some extent on the European Union and Japanese research and technology application initiatives, would pool funds and talent for research and technology demonstration activities. Organizations such as CERF and the Construction Industry Institute (CII) are good examples of the civil engineering profession's and the U.S. design and construction industry's efforts to adopt the lessons of overseas experience by forming such cooperative institutions.

The consortium approach permits a variety of large and small, public and private organizations to join together with limited risk and investment. Private sector companies can still maintain their competitive positions in the marketplace through the development of their own unique product lines or services, based on what they learn in the joint program activities. This minimizes the costs and risks to the companies while maximizing the benefits. This type of organizational structure optimizes contract research and technology demonstration management, and serves as a catalyst for the industry as a whole.

Another transferrable lesson we would extract from our observations is that international cooperation will be increasingly essential for effective innovation in design and construction. The focus in most of our discussion has been on improving the ability of U.S. industry to innovate. We find the continuing experience of European "harmonization," nevertheless, to be a convincing argument that the health of domestic and international design and construction industries will be more closely interrelated in the future. The long service lives of facilities and cyclical nature of investment provide us with relatively fewer occasions to try new ideas, and we should seek to capture each opportunity that arises, regardless of where it is located.

Pursuing effectively the opportunities for new technology, in all modes of infrastructure, no doubt exceeds the resources of any one nation. Targeting efforts on specific technologies or applications or both enhances the likelihood of success, while cooperation enables broader coverage and assures that everyone can learn from each new system. Cooperation would also encourage standardization and enlarge the market for new technology, and thereby offer more attractive commercial rewards to motivate private sector involvement in providing traditionally public infrastructure. In addition, international cooperation will encourage cultural diversity in the search for innovation, and that diversity will enhance the likelihood of new ideas.

Private U.S. companies have effectively become more international, through investments in foreign firms and active pursuit of major overseas project development. These firms inevitably find that their activities are influenced by government policies and, in turn, often have consequences in which governments take an interest. If the U.S. design and construction industry is to gain the greatest benefit from international coopera-  tion, it seems to us that governments must take an active role in expanding their partnerships across borders. Export promotion and development aid agencies can lay the groundwork in the absence of other international agreements to hurdle the obstacles to international applications of new technology. As we have discussed in the domestic setting, such obstacles today include formal regulatory restrictions as well as less formal standards of practice. Obstruction of innovation, whether on protectionist grounds or otherwise, imposes costs on owners and users in the nations that lag behind the advances of newer technology.

RE-ENGINEERING AND RETREADING

Mike Hammer and Jim Champy became celebrities with their advice on how major corporations should "re-engineer" themselves to meet the challenges of modern global business. Their term is highly evocative of the often dramatic ways their advice could alter how these large corporations do business. Companies took a new look at what they were producing and who were their real customers, and made management changes to cut costs and boost productivity and quality. Literally thousands of employees, many from the ranks of middle management, found themselves "re-engineered" into opportunities to find new careers.

It is this latter consequence that fostered resistance in many companies to the ideas Messrs. Champy and Hammer espoused. The authors learned a lesson and have begun to prescribe solutions that include more productive redeployment of personnel resources.

In broad terms, we are talking about re-engineering the U.S. design and construction industry, a relatively labor-intensive collection of enterprises. The industry's many professionals, craftsmen, and skilled workers are the real key to innovation, because they are both an important source of new ideas and the essential participants in putting these new ideas into practice. The experience of re-engineering in other fields represents a final lesson that we believe is important here. We must maintain a highly productive labor force if we are to sustain innovation over the long term. We must also remain mindful of the many small firms that account for the major share of the design and construction industry.

We have seen signs that the lesson may have been missed in both Japan and Europe. Economic recessions in both areas in recent years have hurt business for design and construction. The downturn has come at a time of significant changes in how the industries operate, many of which we have already discussed. A shift is occurring, particularly in Japan, in the relationship between individual workers and their employers. A traditional long-term commitment has been weakened. In Japan this commitment extended to the expectation that a relationship, once established, would continue for the worker's lifetime. With this change, what might have been a problem of inefficiency within an organization becomes a matter of concern in public policy, as potentially productive workers seek new livelihoods.

Providing avenues for workers of all levels to change and grow with the industry will be a necessary element of any realistic strategy for enhancing innovation. That will call for continuing education, retraining, and refocusing, giving some workers essentially new careers and assuring that all workers remain productive throughout their working lives.

ENDNOTES

1. U.S. Department of Commerce and Census Bureau data, as reported in *Building for Tomorrow: Global Enterprise and the U.S. Construction Industry*, National Research Council, Washington, DC: National Academy Press, 1988.

2. Ron Yates. "Not Level with Competition." *Chicago Tribune*, Nov. 9, 1992.

3. Civil Engineering Research Foundation European Research Task Force. *Constructed and Civil Infrastructure Systems R&D: A European Perspective.* Washington, DC: Civil Engineering Research Foundation, 1994; Civil Engineering Research Foundation Japanese Research Task Force. *Transferring Research Into Practice: Lessons from Japan's Construction Industry.* Washington, DC: Civil Engineering Research Foundation, 1991; Japanese Technology Evaluation Center (JTEC). JTEC Panel Report on Construction Technologies in Japan. Baltimore: Loyola College, 1991; and 1992 U.S. Tour of European Concrete Highways (U.S. TECH). *Report on the 1992 U.S. Tour of European Concrete Highways.* Washington, DC: Federal Highway Administration, 1992.

4. "Japan Plans to Double Spending for R&D by 2000, Ministry Says." *Bureau of National Affairs* Aug. 2, 1995. Washington, DC: The Bureau of National Affairs, Inc., 1995.

5. Civil Engineering Research Foundation Japanese Research Task Force. *Transferring Research Into Practice: Lessons from Japan's Construction Industry,* p. 26.

Chapter 6

THE INNOVATION PUZZLE

Now we come to the hard part. There certainly have been successes with past design and construction innovations and there definitely are many opportunities and needs for innovation to continue in the future. Every field has its problems with innovation, so what are we so concerned about, you might ask? Whether you are an engineer, a scientist, a researcher, or merely someone with a passion for experimenting, sooner or later you learn that very few great ideas ever make it to the marketplace. There are serious impediments

Impediments to Innovation Are Especially Severe in Design and Construction

General impediments

- Tort liability, threat of litigation, and high cost of insurance

Structural characteristics

- No single government agency in total charge of construction
- Large numbers of small firms, operating in limited markets
- Multitude of regulatory codes and standards
- Long service lives of facilities and their components
- Cyclical downturns in design and construction markets

Cultural factors

- Procurement policies, particularly in the public sector, that emphasize lowest initial cost rather than best performance
- Divided and often adversarial views of labor and crafts participants
- Reluctance of private firms to invest in experimentation, research and development for longer term profit
- Strong reliance on past experience, attitude of "If it ain't broke, don't fix it!"
- Pervasive public attitude regarding construction, "Not in my backyard!"

that one must contend with. After years of hard work, tests, and evaluations, there are no guarantees that your product will be accepted or used.

But many people would assert that problems are worse in the United States, that innovation in some segments of U.S. industry has become an endangered species. Studies show that the introduction of new engineering products into the U.S. marketplace can take 10 to 15 years, while Japan and other countries are able to do it in half that time. Congressman George Brown of California, a former Chairman of the House Committee on Science, Space and Technology, said that "America has put up so many barriers to innovation, it is self-destructing."

It can be risky to try something new, of course, however beneficial the innovation might be to industry or to society as a whole. There are investments of money and time to be lost if the new idea does not work out. Sometimes, there is also the possibility that people could be hurt or even killed. Prudence is essential. But we may have carried prudence too far. We believe that it is at least as difficult to achieve innovation in design and construction as in any other U.S. industries, and possibly more so. Some of the reasons stem from the industry's unique characteristics, but others are simply extreme cases of problems everyone faces. We will start with the latter.

A KINGDOM OF TORTS

In the United States today, concern about legal liability—the tort system—has created what many of us feel is a crisis. Estimates are that the cost of the U.S. tort system as a percentage of GDP is more than triple that in other industrial countries; it was estimated at 2.6 percent of GDP in 1986, and continues to grow. According to *Forbes* magazine, tort claims cost the country $117 billion in 1987 to *Forbes* magazine, tort claims cost the country $117 billion in 1987 with indirect costs at least as high as the direct costs.[1] Estimates put the 1992 cost at well over $300 billion.[2] The *Forbes* article asserted that "the tort system's direct costs impose a burden in the U.S. five times that in the U.K., and almost seven times the level in Japan."[1]

Data and solid analyses on this issue are weak for design and construction, but a survey by the Conference Board showed that liability concerns in other industries led 36 percent of the respondents to discontinue existing products, while 30 percent decided not to introduce new products,

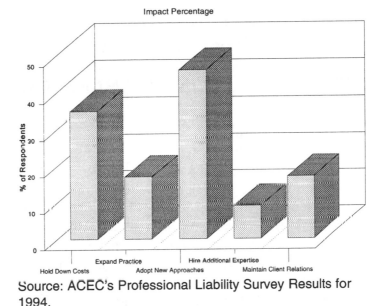

Impact Percentage

Source: ACEC's Professional Liability Survey Results for 1994.

and 21 percent discontinued product research entirely![3] A survey by the American Consulting Engineers Council supported the impression that liability threats are suppressing innovation in design and construction. Design engineers report they feel constrained in their ability to hold down costs for their clients, expand their practices, try more innovative approaches, hire additional expertise, and maintain a desirable relationship with their clients.[4]

While insurance is available, litigation has pushed up the costs. Many design firms pay more for their professional liability coverage, as a percentage of their annual sales, than they show as profit at the end of the financial year! Evidence suggests this situation has little to do with the levels of technical risk.[5]

Because this liability issue affects many segments of U.S. industry, ongoing discussion in Washington and elsewhere could lead to tort system reform. There is a temptation then to say that the design and construction industry should simply participate in defining what this reform should be and urge the nation's policymakers to make the reforms. However, we disagree. When we hear people make that argument, we are reminded of the old joke of the person who was told "Cheer up! Things could be worse!" He cheered up, and things did get worse. The design and construction industry must deal with its own problems, as we discuss below.

STRUCTURAL PROBLEMS MAKE IT WORSE

Even without the tort liability problems, the design and construction industry has a number of structural characteristics that make innovation especially difficult compared to other industries. For a start, we have a regulatory system in the United States, generally applied at the level of local governments, developed to control the hazards that construction can pose. Getting a new product or design accepted requires, in principle, that each of the nation's more than 10,000 local jurisdictions be satisfied that public safety and health will not be compromised. Many businesses and government leaders believe that government regulations have gone too far and we are stifling the architect and engineer who are trying to introduce innovation into their designs. However, as pointed out by University of Maryland Professor of Architecture, Roger K. Lewis, it is important to remember that, "regulation, claimed by many business owners and politicians to have reached unacceptable levels, is like cholesterol. There's the bad kind and the good kind." He goes on to say, "...there are plenty of bad regulations in need of modification or elimination, but the idea of properly formulated and fairly applied regulations is indispensable to a civilized society."[6]

There are national-level groups that consider building code changes in response to new technology or better understanding of the hazards to be avoided (for example, how fires spread in tall buildings or the nature of forces that an earthquake places on a bridge). These groups endorse changes and publish "model codes" that are made available to anyone who wants to adopt them. But it is up to the responsible authorities in local jurisdictions to officially adopt changes that are endorsed in this manner. Some juris-

Courtesy of Roger K. Lewis, Fellow American Institute of Architects (FAIA).

dictions are faster to respond than others. Also, local jurisdictions do not always adopt the exact language of the model codes and may impose local amendments. Even further, explicit code approval is only one hurdle in local jurisdiction acceptance—innovators must generally convince the local building official that a new product or design configuration meets the *intent* of the code.

We are beginning to see some changes in the system. A national model plumbing code has been approved and code authorities are optimistic that a national model building code may be developed by the year 2000. The three model building codes have established the National Evaluation Service, Inc. (NES), whose role is to review technologies and products to determine their suitability vis-a-vis code requirements.[7] These groups have been working closely with national groups and government agencies to enhance their evaluation service and make it more receptive to innovation. Despite these promising developments, gaining product acceptance remains a huge challenge in design and construction.

The problem is not restricted to products. New design methods can also encounter resistance. Enhanced understanding of materials' behavior has supported development of "limit-state" methods that maintained safety but increased efficiency in structural elements. While design professionals were confident of the method's value, it took more than a decade for it to be accepted under standard building codes.

In addition, the opportunities for innovation are rarer in the design and construction industry than many other fields. These opportunities arise primarily when there is new construction, replacement or substantial upgrading of an existing facility, or during maintenance. This is one of the reasons that large projects being developed in Japan and Europe have been such valuable test-beds for innovation. Unlike consumer electronics, agricultural goods, drugs, or many other industrial or consumer products for which product lives measure a few years or less, most buildings and infrastructure facilities stay in service for decades. These long service lives, generally considered an important characteristic of good performance, impede innovation.

The typically cyclical nature of demand for construction makes the situation even worse. The downturns both reduce the opportunities to try new ideas and increase the financial pressures not to take chances on the projects that are undertaken.

IF IT AIN'T BROKE...

As if all this were not sufficiently discouraging, we have developed over the years a series of operating practices in U.S. design and construction that make it even harder for people to accept new ideas. In large measure, these practices add impediments that have become a part of the culture of the industry.

It starts with the way owners and users procure design and construction services. The process, particularly in the public sector, often emphasizes lowest cost rather than best performance. In fact, we often do not really know how to judge what better performance

would be in terms of cost and value over the years that a facility is designed to serve its users. So we specify what we think the minimum acceptable characteristics of the facility should be and buy it from whoever claims to be able to deliver at the lowest initial cost. Even in hiring designers, where price competition is generally thought to be contrary to both the public's best interest and professional ethics, the pressure to deliver a product at the lowest cost, without regard to the product's characteristics, pervades much of contemporary practice. Innovation should be viewed and treated as a "value added" in design and construction and not as an extra cost.

In addition, the final customer for the industry's products, the people and businesses that will use a facility, often have little say in what goes on in design and construction. Real estate developers, government agencies, and other intermediaries develop the functional program and specifications, hire the designers and builders, and deliver a finished facility to the user. If the facility's performance fails to meet these users' needs, particularly years after construction is finished, there is often nobody to respond to complaints. It is as though the manufacturer of each car or television we buy goes out of business within a year of our purchase! There is also little opportunity for designers and constructors to build a business on the repetition of sales to satisfied users.

Of course, the user does not know what goes on during production. Much of the work in design and construction is done by craftsmen, unionized labor, and professionals who practice within tightly defined specialty areas and often guard their "turf" carefully. The views of electrical and sheet-metal workers, carpenters and steel fabricators, architects and civil engineers, and other groups often differ on the merits of new technology. Discussions of how innovations can be made may turn quickly adversarial if one group feels its livelihood may be threatened or that advantages are to be gained by seizing the initiative.

Thin profit margins from least-cost bidding and distance from the user go far to explain why large firms in the design and construction industry are reluctant to undertake possibly costly development of new products or procedures. As is the case in other industries as well, those large firms that are publicly-held corporations are driven by the need to maintain short-term financial results so that their share prices remain high. The management of such firms often find it difficult to justify research that will yield profits in the longer term, if at all. Even those firms that are privately held cannot count on a steady demand to cover their development costs because the industry is so tied to business cycles. In addition, the many smaller firms in the design and construction industry simply do not have sufficient resources to support substantial research and development efforts. For all of these firms, small and large, the prudent business practice in the U.S. seems to be to wait for someone else to introduce innovation and then be quick to learn about it. In the past, companies could at least look to the federal government to support needed R&D. But current trends in the U.S. to cut back on federal R&D investments have left the industry at the mercy of more innovative foreign competition as discussed in Chapter 5.

In addition, the industry has a long history, strong traditions, and a heavy reliance on past experience as a basis for current practice. Many of the participants learn their trade or profession through a process involving an apprenticeship that instills knowledge and respect for the practices that have been proven through long experience to be effective in delivering a safe, durable product. It is hardly surprising then that the old expression "If it ain't broke, don't fix it!" is frequently heard in the design and construction industry. The legal and financial liability that may be incurred by trying something new is another strongly conservative influence.

Finally, the public's attitude toward design and construction has shifted in recent years increasingly toward one of distrust and intolerance of the uncertainties that accompany any effort to innovate. The liability crisis we discussed at the chapter's opening reflects this public attitude, but in design and construction the "Not in my backyard" response has become so widely encountered that its acronym, "NIMBY," has entered the language even in countries where English is not the native tongue.

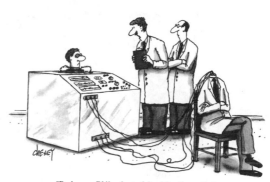

"Bad news, Phil—due to federal funding cutbacks, we can't afford to put your head back on."

Drawing by Cheney; © 1995
The New Yorker Magazine, Inc.

THE PUZZLE OF INNOVATING

We earlier posed the question of whether the U.S. has the incentives in place and an environment for innovation that is needed by the design and construction industry to overcome the barriers and meet the challenges of the coming decades. For the reasons we have cited so far in this chapter, we conclude that the answer is a fairly definite "No!" Each of the barriers to innovation that we have described represents another of the pieces in what we have called the innovation puzzle. In the next chapter we want to make our suggestions of how the puzzle can be put together, but let us now be more explicit about what we think the pieces are.

As we have said, it really is like putting together a jigsaw puzzle, with different people pursuing different strategies. Some people look first for the edges, assembling the frame and then filling it in.

Others look for particular colors or recognizable shapes that they can begin to assemble into larger blocks that must then be placed in their proper positions. Still others are best at looking at the fine details of the shapes of individual pieces, building outward from the first piece they select, in whatever direction the interlocking pieces they find take them.

There is always a danger of being misled by this kind of analogy, but let us push this one a little further. In our puzzle, the edges are like the structure of the design and construction industry, and they are not really fixed. The structure continues to change as U.S. firms form international alliances, adopt new business strategies, and respond to changing markets. In our puzzle, the picture is like the pattern of technology used in the industry, and we do not know exactly what that picture will be when we finish the assembly; it may even change while we work. In our puzzle, the people working to put the puzzle together are like members of our family when we get together for an evening; each one with slightly different motivations, style, and relationships with the others at the table; and each one is certainly likely to see the picture a little differently. But we are all working together toward a common goal.

That is why we find the puzzle analogy most useful. It helps us to describe our goal in understandable terms. In the next chapter we will make some suggestions about how we can change the environment and improve the incentives for innovation in U.S. design and construction. Before we do that, let us look a little more closely at the pieces of the puzzle.

STARTING WITH THE EDGES—ENTRY POINTS FOR INNOVATION

In putting a puzzle together, we like to start looking for edge pieces. They give us some structure, and then guide us into the picture. For us, these are the entry points for new ideas.

Many people describe the process through which buildings or infrastructure are developed and used in terms of five principal stages: (1) planning, (2) design, (3) construction, (4) operations and mainte-

nance, and (5) renewal, reuse, or demolition. In fact, there are many more de- tailed and dis- tinct steps taken over the many years of a facil- ity's service life- time, but these

five are convenient to describe the progression from the initial idea to development of a building or infrastructure, through years of use to obsolescence and beyond. Innovation can enter the process at any stage, as a result of the interaction of stakeholders, their underlying demands and needs, their response to current performance of their buildings and infrastructure, and the technologies available. We have

Stages in the Facility Development Process

- Planning
- Design
- Construction
- Operations & maintenance
- Renewal, reuse, or demolition

already discussed each of these four sets of influencing factors, in previous chapters, but will now put them into context.

In the absence of any major delays, the first three stages of what we call the *facility development process* are typically accomplished over a period of one to three years, although, large, controversial facilities, like Boston's Third Harbor Tunnel or the Los Angeles rail transit system, may take decades to reach the construction stage. Operations and maintenance continue throughout the facility's service life, almost always several decades. With renewal or reuse, the service life can extend to centuries.

In any case, each stage of the development process holds new opportunities for innovation. We have tried to draw a picture of the process below, illustrating the needs and opportunities for new technology to occur all along the way. These opportunities depend, however, on the particular and differing points of view each of the parties to the process (i.e., stakeholders) may have regarding the demands or needs that give rise to consideration of a new technology (e.g., materials, products, procedures); the costs and

Factors Influencing the Adoption of Innovation

- Interaction of stakeholders in the design and construction process
- Underlying demands and needs
- Responses to current building and infrastructure performance
- Technologies available

benefits of that new technology; or the performance of existing facilities and systems (e.g., their output, waste or pollution, and asset value), that gives rise to new demands and needs.

For example, an improved material (i.e., a new technology) may motivate the substitution of that material for another. The substitution will lead in turn to the loss of sales for some vendors and the

Facility Development Process

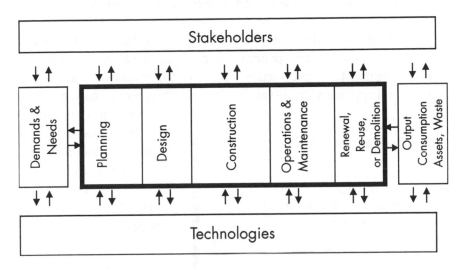

loss of work for those who deal with the replaced material, while others may gain. Some participants in the process may oppose new technologies that other participants favor, because the costs and benefits are (or appear to be) distributed disproportionately or simply because the new idea has been proposed by someone else. The uncertainty of future events makes it difficult for any of the participants to judge what the ultimate value of the potential innovation is, from their perspective, and some may lack the information or insight to make that judgement. These uncertainties and the many participants in the process, to a large extent, account for the generally tortuous path new products and processes face on the road to becoming innovations.

SEEING COLORS AND PATTERNS—TECHNOLOGY APPLICATIONS AND NEW PRACTICES

Once we have started from the edges of our jigsaw puzzle, we begin to look for pieces with similar colors or patterns that might fit into the framework. These are the applications of new technology—the innovations themselves. They, and the facilities and services in which they are imbedded, are the picture in our puzzle.

People who study innovation on a regular basis often describe it as a process with four primary stages: (a) research or discovery, (b) development, (c) production, application, or demonstration, and (d) technology transfer, dissemination, or commercialization. These stages do not necessarily follow a strict and consistent sequence, although research or discovery generally comes early, if not first, in the process.

> **Four Stages of Innovation**
>
> * Research or discovery
> * Development
> * Production, application, or demonstration
> * Technology transfer, dissemination, or commercialization
>
> **...may not follow a strict sequence.**

The difference between discovery and research, in this case, is a matter of whether a conscious effort was made to find the particular new idea. Research is a deliberate undertaking, while serendipitous discovery is what may happen along the way.

Sometimes innovations result when technology is transferred from one area of application to another. For example, medical technologists have learned how to use magnetic resonance of atoms and molecules, a phenomenon once considered an esoteric oddity in physics, to detect differences in living tissues, and thereby develop 3-dimensional images of structures within the human body. This technology is being adapted, in turn, to finding flaws in bridge structures. Through a similar process, devices originally developed to monitor stresses and fluid flow in large hydraulic structures—e.g., dams and electric-power generating plants—are being adapted for use in the cardiovascular system of humans. In each case, we might argue that technology transfer or dissemination started the second-round innovation process, followed by research to solve the problems of adaptation.

Whatever its source, the new idea or technology must typically be developed further to suit its potential applications. Production processes must be developed. Materials must be selected. Connections to other products and systems must be perfected. Only after many problems have been solved is the new process or product ready for trial production, demonstration, or application in the field. It is not surprising that many potential innovations return to further development following first demonstrations. Many other potential innovations die without making it to that stage.

The process can easily break down due to lack of funding in the demonstration phase. Promising technologies become mired in the test phase when the larger sums of money required for a full-scale demonstration are not available. We think the name sometimes given to this part of the proc ess, the "innovation valley of death," is particularly apt.

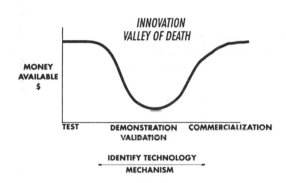

Even if a new idea makes it through this valley, it still may not succeed. The idea must be commercialized by reaching the broad markets that hold the niches where this innovation can fit in. Good ideas can spread relatively quickly in the design and construction industry, through professional contacts, industry news media, and the work of marketeers, but they still may encounter the impediments of regulatory restrictions, labor resistance, or, when there is a cyclical downturn in construction activity, a simple lack of opportunity to apply the needed product.

PUZZLERS MAKING DECISIONS—STAKEHOLDERS IN THE PROCESS

In whatever order they occur, the stages of innovation progress through the actions of particular individuals and groups who

Many Stakeholder/Participants in the Innovation Process

- **Owners**
  Private companies and individuals
  Government agencies
  Regulated utilities
  Stock- and bond-holders
  Mortgage and other lenders
  Insurers
  Others
- **Designers**
  Engineers
  Architects
  Landscape architects
  Financial analysts
  Regulatory specialists
  Others
- **Manufacturers**
  Equipment
  Building products
  Material suppliers
  Fabricators
  Vendors and distributors
  Others
- **Constructors and operators**
  General contractors
  Specialist subcontractors
  Craftsmen and labor
  Maintenance staffs
  Facilities managers
  Others
- **Users**
  End users (e.g., commuters, water customers)
  Intermediate users (e.g., freight forwarders, food products manufacturers)
  Neighbors and other impacted parties
  Others
- **Others**
  Administrative personnel
  Enforcement officials
  Others

participate in the facility-development process. These participants or stakeholders are the people who must put together the puzzle. We tend to group these participants generally into five broad categories: owners, designers, manufacturers, constructors and operators, and users.

In fact, there are often large numbers of people and organizations involved with any particular building or element of infrastructure. The owner, designer, and constructor comprise the major participants, but each of these three is in fact a complex group of individuals and organizations that must work together to accomplish the aim of a completed project. The users are often not explicitly included in much of the process, but they are the underlying reason that everyone else is involved.

The initiator and central figure in the

process, the facility owner, is responsible for stating his or her needs and employing appropriate resources to meet those needs. The user is normally the basis for determining needs, but in many cases the owner and user are different and possibly separate organizations.

For much of the nation's infrastructure, for example, government agencies and regulated utility companies act as proxies for the millions of commuters, water and electricity users, truck drivers, and others who rely on the infrastructure for their safety and livelihoods. The interrelationship of the owner with the other participants in the facility development process, as shown below, can be fairly complex, which contributes to the difficulty of introducing innovation into the process.

The owner's interest in the facility is, in principle, long term. The owner will represent, work with, and employ the other participants throughout the facility's lifetime, from initial planning to final demolition or reuse. The actual identity of the owner may change when buildings are sold or government agencies are restructured, but the owner's role and responsibilities continue unbroken.

The planning, design, and construction stages of facility development involve a great deal of management complexity. In these early stages, the owner typically enters into contractual relationships with designers and constructors. The contracts—legal documents describing what is essentially an agreement among the participants to undertake a stage of the facility development—in principle contain a clear statement of all of the owner's requirements that are to be met during the work leading to the completed project. The contract also states the amount of money the owner agrees to pay for having these requirements met within the design and construction of a facility. In practice, the contract is often unclear or requirements shift during project development. In these cases, contract changes, negotiations, and sometimes disputes contribute to the management complexity.

While the management complexity represents a challenge for participants at various steps in the facility development process, it also represents a source of opportunities for innovation. Each of the decisions that the participant makes along the way is another opportunity.

Interrelationships of Stakeholders in the Facility Development Process

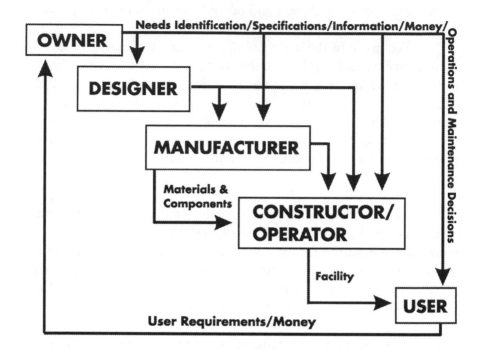

Innovation then may be visualized as new ideas entering the process at various levels and being guided by the actions of stakeholders seeking the best path to the marketplace. Even if the ideas fit well, the stakeholders may not find the other pieces needed to make the fit. Each of the industry stakeholders helps find pieces of the puzzle and tries to put them in place until the puzzle has been completed and the path to the marketplace defined.

WE HAVE MET THE ENEMY...

Pogo the 'Possum, the cartoon creation of the late Walt Kelly, once observed, "We have met the enemy, and he is us!" That is the way it is with innovation in the design and construction industry. As we advance into the 21st century, participants in the U.S. design and

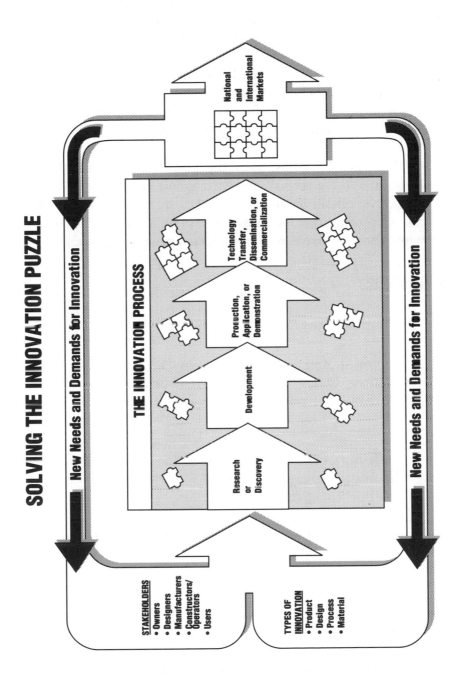

construction industry must learn to work more effectively together. With a growing world population, unquestionably finite natural resources, and an increasingly diverse set of objectives representing our aspirations for ourselves and future generations, we face a pressing challenge: How can we continue to enjoy a high quality of life while assuring that everyone, now and into the future, has the same or better opportunity we seek for ourselves? The term "sustainable development" is often used to capture the concept.

We argue that innovation is the only way we can meet the challenge. Because there are so many participants in the process through which innovation can occur and the stakes are often so high, we have been slower to innovate than have some other industries. It is time for changes in how we do business; it is time to assemble the puzzle.

ENDNOTES

1. *Forbes,* February 17, 1992.

2. Harvey M. Bernstein. "Tort Liability: Limiting U.S. Innovation." *Civil Engineering*, November 1992.

3. Ibid.

4. American Consulting Engineers Council. *Liability Report: Professional Liability Survey Results for 1994.* Cleveland: Penton Publishing, Inc., 1994.

5. For example, refer to the analysis by Michael D. Green, "Tort Law, Deterrence and Innovation: Too Much or Too Little" in *The Role of Public Agencies in Fostering New Technology and Innovation in Building*, D.R. Dibner and A.C. Lemer, editors. Washington, DC: National Academy Press, 1992, pp. 101-124.

6. Roger K. Lewis. "Architects Are in a Zone of Their Own Over the Abundance of Regulations." *Washington Post*, February 18, 1995, p. F3.

7. The three model building codes include: Building Officials and Code Administrators International, Inc. (BOCA), International Conference of Building Officials (ICBO), and Southern Building Code Congress International, Inc. (SBCCI).

Chapter 7

SUSTAINING INNOVATION

The satisfying part of putting together a jigsaw puzzle is when you can stand back, after the last piece is put in, and look at the whole picture. Everyone who worked on it can share a sense of accomplishment and for the first time see how everything fits together.

This is where our analogy to innovation in design and construction is the weakest. With innovation, we will never really see the whole picture. Certainly we can achieve success on a project-by-project basis, bringing a particular new idea to market, or solving a particular problem. We can take satisfaction from each piece that falls into place along the way. But there are so many opportunities for innovation, so many problems to be solved, so many stakeholders and complexities in the process, that we can never, as individuals, see the whole puzzle finished.

But this is good! The process of innovation in design and construction should be continuous. We can make progress, and our aim here has been to help speed that progress. We can focus our efforts on how to refine the innovation process and unite the various stakeholders in the design and construction industry. As more and more of the puzzle is put together, the better we may become at the tasks of assembly. Each innovation will benefit the industry, the nation, and the world.

U.S. industry woke up several years ago to what some people have termed the threat of a Japanese "invasion." Japanese auto makers, electronics manufacturers, and others were seemingly flooding the U.S. market with very high quality and reasonably priced goods. Government policy makers in Washington, D.C. watched the nation's trade deficit skyrocket as U.S. consumers flocked to buy these attractive imports. In design and construction, the threat appeared in the form of a declining market share for U.S. business competing for work abroad and in a growing level of investment by Japanese companies in the United States.

The flood has slowed somewhat, but we still have lessons to learn from the experience, as our own observations have made plain. One of those lessons is reflected in the Japanese word *keizen*. Generally translated as "continuous improvement," *keizen* refers to a philosophy of always searching for ways to do a better job at whatever you are doing. If everyone in the organization adopts this philosophy, a steady stream of small efficiencies add up to large advances in quality and, in the business setting, in the competitive position as well. For a brief period, *keizen* and its English equivalents became popular in the U.S. business press as we debated how we could learn to compete against the Japanese competition onslaught.

For the design and construction industry, we are aiming toward an expansion of this philosophy of *keizen*. We must aspire to *sustaining innovation*.

SEEKING SUSTAINABLE INNOVATION

We mentioned the term "sustainability" earlier. It is one of those concepts that sounds very good in the abstract but turns out to be stubbornly resistant to precise translation into practical terms. Most people begin with the basic statement developed by the U.N.-sponsored World Commission on Environment and Development. The commission's report on *Our Common Future*, prepared under the chairmanship of Gro Harlem Brundtland, Prime Minister of Norway, and published in 1987, is widely credited with raising the term to its present importance. The Brundtland Report said that "Sustainable development is development that meets the needs of the present without compromising the ability of future generations to meet their own needs."

Several factors will influence how this definition is translated into practical guidance for the design and construction industry. First, we have to recognize that human needs are growing: our population is increasing; most people aspire to have the higher living standards they see that others have achieved; and there seems to be a steady increase in the aspirations of those who now live at very high standards. We are unlikely as a species to suppress our desire for what we believe to be a better life. Second, we have to acknowledge that there are absolute limits on the availability of minerals, air, water, scenic locations, and other elements of our terrestrial home— "Spaceship Earth" as Buckminster Fuller termed it—that we

Although these windmills possess little resemblance to their Dutch predecessors, they are one of the many technologies being tested as a sustainable energy resource for the future. *Courtesy of Paul Grabhorn.*

tend to call "resources" when we find a use for them. Technology may find new sources, increase our efficiency in using these resources, or find substitutes for those that are in short supply, but ultimately each element can be "used up." When this happens, future generations will have lost that resource. There is an obvious potential conflict between these two propositions, that growing human needs should be met and that limited resources must be conserved. Resolving this conflict is a challenge the design and construction industry must take on.

Throughout 1995, we have both been working with a multinational group of design and construction industry leaders and government officials from Europe, Japan, and the United States on the planning for a symposium on *Engineering and Construction for Sustainable Development in the 21st Century*. That group, grappling with the challenge, concluded that sustainable development can be met only through substantial innovation and, of course, we agree.

There can be little question that design and construction have something to do with sustainability. On the one hand, our buildings, infrastructures, and other elements of the built environment are assets that can be passed down to provide shelter and support for future generations. On the other hand, these assets can deteriorate, wear out, and become obsolete, and thereby represent a waste of resources available to these same future generations.

We would assert that innovation in design and construction is crucial to achieving sustainability in the broad sense that the global discussion signifies. Innovation will both enhance the value of our infrastructure and buildings as assets to be passed from one generation to another and reduce the wastes associated with our current methods of developing and managing those assets. But even further, innovation itself must be sustainable if we are to meet the "needs of the present without compromising the ability of future generations to meet their own needs." These needs are at best poorly understood today, and our perceptions of them are likely to change as we learn more in the future.

The innovation we are talking about is not just stronger concrete and faster computers. We are targeting directly the services the design and construction industry ultimately provides, rather than the components and stages of production that go into providing those services. There must be stronger linkages to these services from planning and design through construction and maintenance. Automated processes and advanced materials have a role because they improve services or enhance efficiency or both, but they are not ends in themselves. Environmental and social considerations, still viewed largely as constraints on many segments of the design and construction industry, will increasingly be built into the process as real costs of production or components of services delivered.

This focus on services will give everyone a stake in the industry's final product. By eliminating the boundaries between stages in the production process, the improvement of production becomes a matter of importance to all participants. Working together makes more sense when everyone stands to gain. When everyone shares in the benefits of each improvement, the innovation process can become continuous, each improvement in design and construction leading to new opportunities, in the search for continuous improvement. In the end, we believe it is the process of working together that is crucial to sustaining innovation.

Some people argue in favor of basic change in the way the design and construction industry views itself and what its purpose is. We have found a broad consensus emerging among the industry's leadership, that the U.S. design and construction industry must reconfigure itself to achieve the level of coordination we envision. These leaders feel that the industry must have greater "vertical integration," to use the jargon of the management schools. We tend to agree, but understand that many factors are likely to slow the process of change, including the industry's diversity, large numbers of small firms, and traditional separation of design from construction, and construction from operations and maintenance. Not everyone views the prospect of industry integration with favor, and we do not insist that it must occur.

We do insist, however, that greater cooperation and a focus on services is essential for the industry to achieve the level of sustained innovation of which it is capable. Because we see this cooperation and new focus being adopted elsewhere, we also assert that the U. S. design and construction industry must make the change if they are to keep up in the global marketplace.

We propose a strategy for sustaining

> **A Four-Part Strategy for Sustaining Innovation**
>
> - Setting bold but realistic goals
> - Shifting the industry's product to a full lifetime of service
> - Realistically allocating the risks of innovation
> - Enhancing innovation's rewards for all participants

innovation, designed to bring all participants at the table together: industry, academia, and government leaders, owners and users, technologists, financiers, and the many others who have a role in achieving design and construction innovation. Our primary aim is not so much to increase these groups' efforts as to encourage common direction in those efforts. Our strategy has four key elements: setting bold but realistic goals, shifting the industry's product to a full lifetime of service, realistically allocating the risks of innovation, and enhancing innovation's rewards for all participants.

SETTING GOALS FOR IMPROVEMENT

We start with goals. Perhaps "vision" would be a better term. 19th-century architect-planner Daniel Burnham advised "Make no little plans...", and we believe a bold vision is needed now. We cannot expect that an earthquake or other disaster will foster the crisis mentality that we have seen can lead to rapid innovation such as what occurred in the immediate aftermath of the Northridge earthquake. We certainly would not wish to experience the suffering and loss that a major disaster would entail. But we do seem to need a similarly strong motivation to act. Here again we can draw on the industry's leadership.

White House Task Force Proposed Industry Targets for Innovation

- 50 % shorter project delivery time
- 50 % lower operations/maintenance costs
- 30 % greater facility comfort & productivity
- 50 % fewer building-related illnesses and accidents
- 50 % fewer job-related illness and accidents for construction workers
- 50 % less resource waste and pollution
- 50 % greater durability and flexibility

Source: Civil Engineering Research Foundation. *National Construction Goals: A Construction Industry Perspective*. Washington, DC: Civil Engineering Research Foundation, 1995.

In 1994, the White House invited a number of the industry's leaders to participate in a series of meetings held under the auspices of the Subcommittee on Construction and Building within the National Science and Technology Council, to discuss how federal policy might be made more supportive of the design and construction industry. These leaders endorsed a number of rather dramatic goals for the industry over the coming decade, to lower costs, improve safety, and enhance service to the owners and users of the design and construction industry's products.[1] [See previous box.]

Following these meetings the Civil Engineering Research Foundation conducted an international survey of design and construction industry practitioners, researchers, academics, and government officials, to determine what they thought were the key issues confronting the design and construction community.[2] Respondents from over 20 countries—45 percent from outside the United States—offered their perspectives on what would be achievable within the next one to two decades. Not surprisingly, the results (see graph) fell short of the ambitious targets proposed by the White House group. Effective leaders recognize that setting goals high can

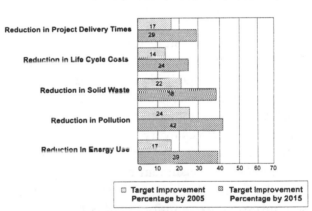

Expectations for Industry Improvements

Reduction in Project Delivery Times — 17, 29

Reduction in Life Cycle Costs — 14, 24

Reduction in Solid Waste — 22, 48

Reduction in Pollution — 24, 42

Reduction in Energy Use — 17, 39

(x-axis: 0 10 20 30 40 50 60 70)

☐ Target Improvement Percentage by 2005   ☒ Target Improvement Percentage by 2015

Source: Civil Engineering Research Foundation. *Engineering and Construction for Sustainable Development in the 21st Century: An International Perspective [Delphi Survey Results].* Washington, DC: Civil Engineering Research Foundation, 1995.

spur people to achieve their best, but one can imagine these survey respondents thinking about their current workloads and budgets.

Setting out targets is an important step, and it is helpful if everyone is aiming at least in the same direction. For this reason, professional and trade groups in the design and construction industry have a role to play in adopting and publicizing innovation goals. Establishing high levels of industry and public awareness of the innovation effort will help to maintain an enabling environment for innovation to occur.

To stay competitive, firms will find greater reasons to be more innovative if they see that their competitors are benefitting from its applications and their customers are more receptive to the life-cycle advantages of innovation. We believe that the customers are in fact a key piece of the puzzle. The people who hire designers and builders and who own and use facilities must insist that the industry set its sights high.

PUTTING YOUR MONEY WHERE YOUR MOUTH IS

Of course, vision alone will not work. To be globally competitive, the construction industry leadership must go further and make the resources available that are needed for the campaign. We think a comparison to the electronics industry is illuminating. That industry's brief history is dominated by the experience that the power of computer chips doubles every 18 months, and cost plummets, leading to new products and upgrading throughout the industry. The steady stream of accomplishment may be interrupted in the future, but for now this "rule" (known as Moore's Law) guides corporate decisions, sets investors' expectations, and is to some extent self perpetuating. Industry leaders like Intel and Motorola devote substantial resources to research, reportedly equivalent to eight to ten percent of their total volume of sales. Their management expects a payoff from these expenditures and they get it, or they fall behind their competitors.

The design and construction industry must find ways to allocate more funding to all aspects of the search for innovation, including technology demonstrations and dissemination as well as research. For all of the reasons already explained, this will not be easy: paper-thin operating margins in construction; many small companies; many competing interests; and more. The industry needs mechanisms to mobilize its resources and more effectively bring them to bear on the search. We propose that change could come in several stages.

First the clients. The federal government, as a major purchaser of design and construction services and the nation's largest facility owner, could set aside a small amount of all construction budgets to serve as "seed money" for innovation. This budget allocation would be analogous to the one percent of federal construction funding that is currently set aside to commission or purchase art for federal facilities. Several states and cities have adopted similar "percent-for-arts" programs, suggesting that the public will support such spending if they can see the payoffs. Industry support could mobilize the Congressional action that would be required to establish the set-aside program.

Arts set-aside funds are generally spent on each project from which they are collected, but we propose that the innovation set-aside funds should be pooled in a "facilities-trust fund." We term it a facilities trust to convey the purpose of enhancing the assets that buildings and infrastructures represent.

Federal corporate income tax credits for research and experimentation are a useful mechanism for encouraging the search for innovation. We certainly support those members of Congress who are pushing to assure that such credits are available. The value of tax credits is limited, however, when institutional forces and competition keep operating margins paper thin, as is the case in much of the design and construction industry. We therefore have gone one step further in calling for a set-aside and trust fund.

Other sources of revenue would flow into the facilities-trust fund. One source would be the premiums paid for federal guarantees of the innovation insurance risk-sharing program as proposed in Chapter 5. Another could be a small surcharge or excise tax on inputs to design, construction, and facility operations and maintenance (including, for example, building materials, labor and professional payrolls, and energy used in space heating). This latter source would be similar to the construction payroll tax used in Sweden to fund research.

All these sources of funds are tied to government action. The private sector should participate as well, but there are few mechanisms to mobilize more than voluntary contributions to the trust fund. Such contributions could be encouraged if they were payable out of gross revenues and exempted from federal income tax, much like a charitable deduction. Owners, as well as design and construction companies might choose to contribute to the trust fund.

A more reliable stream of revenue would be preferable, and could be established if there were a means for collecting an innovation levy on all building projects at the time that building permits are issued. Such a levy, based on estimated project cost, generates about half of the income of the Building Research Association of New Zealand (BRANZ). The levy and BRANZ's work enjoy industry support. New Zealand's more centralized government structure enabled the levy to be put into place much more easily than would be the case in the United States, where local government authorities are generally responsible for issuing building permits. We frankly have not been able to think of an effective way to encourage the thousands of local governments to enact legislation, collect, and efficiently pass along such a levy to a centralized trust fund. Setting up separate trust funds at the state level might bring the concept closer to feasibility, but to some extent would run counter to our effort to consolidate resources.

Another approach to mobilizing private-sector resources might be for the trust fund institution to take equity shares in new technologies, acting as a venture-capital investor. Direct participants in a demonstration project would retain ownership of a majority share of the technology to be demonstrated, but a share of the ownership would be transferred to the corporate pool. The fund would be in some respects similar to a mutual fund that invests in high technology start-up companies. The trust fund would be structured to allow sale or "spin-off" of these equity shares, further distributing the risk and potential financial rewards of design and construction innovations. A secondary market in these public innovation shares might be expected to develop.

Because the trust fund would be used in part as an innovation-risk management tool, only a fraction of the fund's receipts would be spent for research and demonstration activities. These amounts would not have to be very large, however, to represent a substantial increase in the design and construction industry's current commitment to these activities. Raising only two percent of the nation's yearly spending on construction would bring in approximately nine billion dollars. If only one-fourth of that amount could be spent on research, demonstrations, and related activities, it would be almost double the amount the federal government now spends annually on construction-related R&D in the United States and more than six times what the U.S. construction industry currently invests in R&D. With effective management of the risk-sharing program, it seems likely the spending for research and demonstration could go up.

CHANGING TO A LIFETIME PRODUCT

In present practice, it takes a long time from initial planning of a new—or substantially refurbished—facility to when it can be "commissioned," that is, put fully into service. That time is typically 9 to 18 months for conventional housing, for example, while two to five years or more may be needed for a tall skyscraper. The "project delivery time" for a highway or other complex piece of the infrastructure can easily extend to a decade and beyond. A 50 percent reduction in this "project delivery time," as proposed by the White House group of industry leaders, would save millions of dollars in time, value, and financing costs alone.

But suppose we look further ahead. Experience shows that over the two to five decades that a building or element of infrastructure can be expected to provide service, the costs of operations and maintenance will accumulate to an amount that far exceeds the initial costs of the design and construction. For a typical office building, for example, heating and cooling, lighting, maintenance of plumbing, roof repairs, and the various other activities required to keep the building operating properly, account for about 60 percent to 85 percent of the building's total "life cycle" cost, and even more when inflation rates are high. Viewed from another perspective, these continuing operations and maintenance costs, over the course of the building's lifetime, total up to an amount that is several times the initial cost of construction.

The point is that needs and opportunities for design and construction innovation occur over long periods of time but, as we have pointed out, most of the major decisions are made during the initial project delivery period. The designers and constructors, for the most part, finish their task before the real product-service to the economic and social activities of facility users—is delivered. This must change if we are to establish an environment conducive to sustaining innovation.

All stakeholders involved in the delivered product should be held accountable throughout a facility's service life. This will likely mean greater integration in the industry, with "full-service" firms or consortia consisting of many smaller firms offering to design, build, operate, and maintain facilities. Such arrangements are in fact already being used for occasional projects such as toll highways, waste treatment, and electric-power generating plants. Some of the incentive for these arrangements has come from cash-strapped governments seeking to finance infrastructure that can support development. Such terms as BOO (for build-own-operate) and BOT (build-operate-transfer) are used to refer to the management and ownership relationships that are established.

We believe such arrangements are good for the industry and should be expanded. The actions to do this will have to come primarily from facility owners and users. We propose that large corporations or government agencies that are major investors in real estate or infrastructure (e.g., electric utilities) are the most likely to benefit from the change. In making the change, design and construction would become more like other industries, closer to their ultimate customer. As we learned in Europe, companies confident of the quality of their product are able and willing to give a warranty on their work. The willingness of top Japanese and U.S. building constructors to work with their clients for more extended periods to assure that the facility is functioning properly is ample evidence that the change is even compatible with today's practices.

BALANCING RISKS

Of course, if a company is to be held responsible for several decades of the service lifetime of a facility, there will be a greater concentration of business risks than most participants in the design and construction industry today face. Ways are needed to control these risks, to allocate them realistically among stakeholders, and to assure that the risks do not stifle innovation.

Drawing on the lessons of Europe and Japan, we propose that a peer review system should be established for examination of prom ising design and construction concepts that could weigh their poten-

tial benefits against the possible threats to human safety, property, or environmental quality. Design and construction professionals, the insurance industry, financial institu-

tions, and regulatory bodies would join together in this effort, perhaps by forming one or more institutions modeled on the Underwriters Laboratories, the Civil Engineering Research Foundation's Highway Innovative Technology Evaluation Center (HITEC), or Europe's Agrément system. Institutions such as these help to knock down the barriers for getting the innovation to the marketplace.

Developers of new technology would apply for a consensus-based evaluation that would measure how their new product or process performs compared to pre-set criteria, and judge whether it is suitable for full-scale field demonstration. Such demonstrations are crucial. "Proof of concept" is a *sine qua non* for adoption of innovation in any industry, and the design and construction industry is no exception. Indeed it is even more dependent on the demonstration, given the need to protect public safety. Yet as we have discussed, the business and professional risks represented by the U.S. tort liability and financial systems are effective deterrents to many demonstrations.

The consensus-based evaluation would be accepted as a basis for setting insurance rates and granting building code variances or other necessary regulatory clearances, and thereby help to raise the floor of the "innovation valley of death." Such institutions could operate internationally or establish reciprocity or liaison relationships with counterpart national organizations.

SHARING RISK BENEFITS ALL PARTICIPANTS

The costs of operating the assessment programs, including necessary testing, would be paid in part by fees collected from developers seeking analysis of their new technologies. However, a portion of these costs should be shared across the industry, as a recognition of the broad benefits of innovation, by drawing on the "facilities-trust fund."

The consensus-based peer-review institutions would play a role in allocating the research and demonstration spending. Major insurers participating in the technology-evaluation process would be well placed to undertake the pooling and distribution of risk associated with

demonstration projects. Facility owners as well as designers and constructors might seek funding to support demonstrations of new technology. There are many ways to solve a problem, and sharing the risk
can help ensure that a solution is found and everyone benefits.

Facilities that serve federal agencies and those that have been
developed with substantial government support could similarly be
used as a test-bed for new technology that might have broad application. This use of federal assets would be analogous to spending for
national defense and the technological advances that weapons systems development fosters. Local communities seeking to attract economic development and revitalize urban precincts could seek funding
as well, using their revitalization efforts as an opportunity to incorporate new infrastructure technologies and shift development patterns
to use these new technologies most productively.

ACHIEVING REWARDS

Incentives already exist for innovation, of course, including
profits to be made and the psychic rewards that drive many researchers and innovative professionals to be the "paradigm pioneer"
or "paradigm shifter." We have argued, nevertheless, that increased
research and technology application efforts in design and construction are warranted. We must acknowledge that the current levels of
activity are negligible—as we have already noted, annual U.S. spending on design and construction industry research and development,
by most accounts, nearly all from federal sources, exceeds $2 bil-
lion—but the returns could be greater.

As we said earlier, government has another key contribution
to make here, through tax policy. We propose that companies making
expenditures on design and construction research and technology
application should be permitted to treat those expenditures entirely
as expenses for tax purposes. If a research entity cannot use its tax
credits in the year the expenditures are made, the company should be
permitted to carry these credits forward or sell them in much the
same fashion that air pollution credits may now be traded. This latter
incentive could assist potential innovators in the last stages of devel-

opment to overcome the cash shortages that often threaten to prevent demonstration of new technology.

Finally, there must be broader recognition of the value of achievements in design and construction innovation. This recognition must come largely from within the industry, such as the Construction Innovation Forum's NOVA Award and the Civil Engineering Research Foundation's Innovation Award, but external stimulus will help it to develop. Awards programs, perhaps modeled after the government's Malcolm Baldridge Award for quality in industry, the National Medal of Technology for individual technical achievement, or the Presidential Design Awards program for architecture and industrial design should be established to reward at a national scale the achievements of innovators in facilities technology.

The cycle of action for sustaining design and construction innovation.

There should also be regular reporting of the industry's progress in achieving the goals for design and construction improvement. Just as we now have statistics on residential building permits, total value of new construction, and employment in construction, so there should be data collection and reporting of relative project duration, facility age, and other indicators of how well the industry is performing. There are of course many difficulties in developing a set of such indicators, and while much of the data needed are already being collected, other data collection may be required. Government statistics agencies and industry groups will have to work together to design and implement the data collection and reporting system.

PIECING TOGETHER THE PUZZLE

Our strategy will be effective only if the industry takes action, and we have here proposed a number of such actions that we would recommend. (See summary, box below.) In writing, one must start somewhere, and we started with generating enthusiasm and a sense of direction by setting bold but realistic goals. But we believe that action should start on several fronts. The relationship among our four principal points is more a continuing cycle than a linear path from goals to rewards for innovation.

Much of the action must come primarily from within the design and construction industry itself, but all participants in that industry have a role. Owners and users, both government and the private sector, must recognize the value of design and construction innovation and be willing to take the chance for improvements that new technology can offer. They are the source of demand for innovation and will benefit directly from the lower costs and improved service that innovation will bring. Designers, constructors, and operators must recognize that their markets are not static and that they stand to gain by taking leadership in innovation. They are the source of supply for that innovation, and

---

Specific Actions that Can Encourage Innovation in Design and Construction

- Set national goals for improvement and publicize their adoption.
- Establish life-cycle service delivery as the product of the design and construction industry.
- Establish peer-review institutions to review promising new concepts for demonstration and evaluation in field applications.
- Form a facilities-trust fund and industry-based risk-sharing pools, using insurance, marketable securities, and other appropriate financial mechanisms to reduce innovators' share of risk to acceptable levels.
- Use government budget set-asides, excise taxes, and user fees to generate income for the facilities trust fund.
- Use trust funds to support research and demonstration activities on private and public projects.
- Use government facilities as a test-bed explicitly to encourage innovation by demonstrating new technology.
- Work with selected local communities to incorporate innovative design and construction technology into economic development and community revitalization activities.
- Create tax and other incentives that encourage private sector investment in research and new technology development.
- Establish awards programs and other national recognition for achievements in design and construction innovation.
- Establish data and reporting systems to monitor industry accomplishments relative to the adopted goals for improvement.

will profit by increasing their own productivity and the size of their markets. Together, these participants can bring demand and supply together to solve the innovation puzzle.

But we must bring along the other stakeholders as well. To do this, we must not only promote the industry's use of innovation. We must also help the public to understand the benefits as well as the risks of innovation, and assure that the broad community of facility users participates in establishing the appropriate balance between those benefits and risks.

We are enthusiastic about the opportunities, but we all must be realistic as well. The results of innovation will become apparent gradually. The history of technological progress teaches us that the dramatic gains in productivity and quality of life develop over periods of two to four decades following the introduction of major new technologies. If we start now, our children will be the primary beneficiaries of our efforts.

We think this is as it should be. Sustaining innovation in design and construction is an essential element of a sustainable future. If we are to achieve this future, we must set to work now on completing the innovation puzzle.

ENDNOTES

1. Civil Engineering Research Foundation. *National Construction Goals: A Construction Industry Perspective.* Washington, DC: Civil Engineering Research Foundation, 1995.

2. Civil Engineering Research Foundation. *Engineering and Construction for Sustainable Development in the 21st Century: An International Perspective [Delphi Survey Results].* Washington, DC: Civil Engineering Research Foundation, 1995.